01-CTS-4

Develop and Demonstrate Fundamental Basis for Selectors to Improve Activated Sludge Settleability

by:

Donald M.D. Gray (Gabb), Ph.D., P.E., BCEE
Vincent P. De Lange, P.E.
Mark H. Chien

East Bay Municipal Utility District

2006

WERF

The Water Environment Research Foundation, a not-for-profit organization, funds and manages water quality research for its subscribers through a diverse public-private partnership between municipal utilities, corporations, academia, industry, and the federal government. WERF subscribers include municipal and regional water and wastewater utilities, industrial corporations, environmental engineering firms, and others that share a commitment to cost-effective water quality solutions. WERF is dedicated to advancing science and technology addressing water quality issues as they impact water resources, the atmosphere, the lands, and quality of life.

For more information, contact:
Water Environment Research Foundation
635 Slaters Lane, Suite 300
Alexandria, VA 22314-1177
Tel: (703) 684-2470 Fax: (703) 299-0742
www.werf.org werf@werf.org

This report was co-published by the following organization. For nonsubscriber sales information, contact:

IWA Publishing
Alliance House, 12 Caxton Street
London SW1H 0QS, United Kingdom
Tel: +44 (0) 20 7654 5500
Fax: +44 (0) 20 7654 5555
www.iwapublishing.com
publications@iwap.co.uk

Library of Congress Catalog Card Number: 2005925038
IWAP ISBN: 978-1-84339-752-6

East Bay Municipal Utility District, Oakland, CA

The research on which this report is based was developed, in part, by the United States Environmental Protection Agency (EPA) through Cooperative Agreement No. CR-827345-01 with the Water Environment Research Foundation (WERF). However, the views expressed in this document are solely those of East Bay Municipal Utility District and neither EPA nor WERF endorses any products or commercial services mentioned in this publication. This report is a publication of WERF, not EPA. Funds awarded under the Cooperative Agreement cited above were not used for editorial services, reproduction, printing, or distribution.

ACKNOWLEDGMENTS

The authors would like to express their appreciation to the Project Subcommittee who provided valuable guidance and input throughout the course of this study. They also thank Orange County Sanitation District, Veolia Water, Inc. (formerly USFilter/Vivendi), OMI, Inc., and numerous municipalities across the United States that contributed valuable information to the project database. The authors acknowledge the efforts of the East Bay Municipal Utility District staff who assisted with the selector demonstration project, including David Freitas, Kurt Haunschild, Edward McCormick, John Cloak, Jack Lim, Sue Berg, Clyde Pham, Lisa Servande, Amar Sidhu, Steve Savage, Steve Kallal, and Alexander Borys, as well as a number of Environmental Careers Organization interns, including Andrew Gentile, Chanice Harris, Xiaozhou You, Matt Hoeft, Carl Anderson, and Janet Chuang. The authors would also like to thank Dr. David Jenkins for providing invaluable comments during technical review of the draft and final project reports.

Report Preparation

Principal Investigator:
Donald M.D. Gray (Gabb), Ph.D., P.E., BCEE
East Bay Municipal Utility District

Project Team:
Vincent P. De Lange, P.E.
Mark H. Chien
David R. Williams, P.E.
East Bay Municipal Utility District

H. David Stensel, Ph.D., P.E., BCEE
Gang Xin
University of Washington, Seattle

Elliott Wheeler
OMI, Inc.

Somnath Basu
Veolia Water, Inc.

Mark Esquer, P.E.
Michelle Hetherington, P.E.
Orange County Sanitation District

B. Narayanan, Ph.D., P.E.
Carollo Engineers, Inc.

Bob Kemmerle, P.E.
E2 Consulting Engineers, Inc.

Project Subcommittee

Orris E. Albertson, P.E., BCEE
Enviro Enterprises, Inc.

Glen T. Daigger, Ph.D., P.E., BCEE, NAE
CH2M Hill

M. Truett Garrett, Jr., Sc.D., P.E.
Post, Buckley, Schuh, & Jerrigan

Tung Nguyen
Sydney Water Corp.

Water Environment Research Foundation Staff

Director of Research:	Daniel M. Woltering, Ph.D.
Senior Program Director:	Amit Pramanik, Ph.D.

ABSTRACT AND BENEFITS

Abstract:

Although selectors have been widely applied to control filamentous bulking in activated sludge systems, significant variation exists in design and operating practices and the degree of sludge settleability achieved. The goal of this research was to investigate fundamental issues regarding the growth and control of specific filamentous organisms at bench scale, develop an extensive database of selector design and operating data from full-scale facilities, and demonstrate implementation of full-scale, pilot anaerobic selectors at two large wastewater treatment plants. Based on data collected from 44 facilities, this project examines the relationship between various process parameters and settleability control.

This study identifies the most significant process variables affecting settleability control in three distinct plant categories—short-MCRT with anoxic or anaerobic selectors, short-MCRT with aerobic selectors, and long-MCRT—and provides recommended design and operating ranges based on single-variable regression analysis of a large database of full-scale plant data. The project team has incorporated this information into a computerized selector diagnostic tool that may be used to retrieve recommended design and operating ranges from the current study and the literature based on user input.

Benefits:

- Evaluates the role of readily assimilable chemical oxygen demand (raCOD) in the growth and control of *Thiothrix* spp.
- Documents selector performance and operating data from 44 full-scale facilities.
- Evaluates the relationship between various process variables and settleability control.
- Demonstrates implementation of full-scale, pilot anaerobic selectors at two facilities.
- Provides a semi-empirical formula for calculating the "effective" number of selector compartments (N) in a selector zone based on flow conditions and basin geometry when dye study results are not available.
- Ranks selector design and operating parameters based on the influence on settleability for three different plant categories—short-MCRT with anoxic or anaerobic selectors, short-MCRT with aerobic selectors, and long-MCRT.
- Provides recommended design and operating ranges for the most critical process variables in each of the three plant categories.
- Provides a computerized selector diagnostic tool (available on CD-ROM attached to inside of back cover of report) to assist in troubleshooting existing selectors or designing new selectors based on user input and design/operating parameter recommendations from this study and the literature.

Keywords: Selector, filamentous bulking, settleability

TABLE OF CONTENTS

LIST OF TABLES

LIST OF FIGURES

LIST OF ACRONYMS

BOD	biochemical oxygen demand
BNR	biological nutrient removal
CMAS	completely mixed activated sludge
COD	chemical oxygen demand
CSTR	continuous-flow stirred tank reactor
DNA	deoxyribose nucleic acid
DO	dissolved oxygen
DSVI	diluted sludge volume index
EBMUD	East Bay Municipal Utility District
EBPR	enhanced biological phosphorus removal
F/M	food-to-microorganism ratio
GAO	glycogen-accumulating organism
HP	horsepower
HRT	hydraulic residence time
ICZ	initial contact zone
LCFA	long-chain fatty acid
MCRT	mean cell residence time
MLSS	mixed liquor suspended solids
mV	millivolts
MWWTP	Main Wastewater Treatment Plant (EBMUD)
OCSD	Orange County Sanitation District
OUR	oxygen uptake rate
PAO	phosphorus-accumulating organism
PHB	poly-β-hydroxybutyrate
RAS	return activated sludge
raCOD	readily assimilable COD
rRNA	ribosomal ribose nucleic acid
saCOD	slowly assimilable COD
sCOD	soluble COD
SBR	sequencing batch reactor
SCFA	short-chain fatty acid
SSV_{30}	30-min settled sludge volume
SVI	sludge volume index
TAG	triglyceride
UCT	University of Cape Town (South Africa)
UOSA	Upper Occoquan Sewage Authority
VIF	variance inflation factor
WAS	waste activated sludge

EXECUTIVE SUMMARY

ES.1 Key Findings

In brief, this study supports the following conclusions:

- Anoxic selectors do not appear to control filamentous bulking in long-mean cell residence time (MCRT) plants. In fact, the elimination of all anoxic zones may help to control bulking in these plants. Other design/operating parameters, however, were shown to influence activated sludge settleability in long-MCRT plants.
- Aerobic selectors in short-MCRT plants do control filamentous bulking if they are small enough to produce a biochemical oxygen demand (BOD) concentration gradient in the aeration basins.
- Anoxic and anaerobic selectors do control filamentous bulking in short-MCRT plants if the selector volume is large enough and/or the selector mixed liquor suspended solids concentration is high enough. These selector systems do not appear to benefit from a BOD concentration gradient as the aerobic selectors in short-MCRT plants do. Although anaerobic/anoxic selector compartmentalization in these plants appears to improve settleability, this is presumably because of reduced selector short-circuiting.

To make the study findings more readily available to practitioners, the project team prepared a computerized selector diagnostic tool, which is included on a CD-ROM attached to the inside back cover of this report. Documentation for this software application is provided in Appendix F, which explains the simple steps to using the selector diagnostic tool software. This study's findings can be used immediately through this software to help an operator troubleshoot a poorly-performing selector or help an engineer design a better-performing selector.

If the practitioner is interested in how the selector diagnostic tool's guidelines were derived, Chapter 4.0 can be referenced. If the practitioner is interested in an actual demonstration of these guidelines, Chapter 5.0 can be referenced. Chapter 6.0 provides a more detailed summary of the study's findings and conclusions, Chapter 2.0 provides a selector literature review, and Chapter 3.0 provides laboratory study results demonstrating the role of readily assimilable chemical oxygen demand (raCOD) in selector performance. Refer to the discussion on raCOD in Chapter 1.0, Page 1-3.

As shown in Chapter 5.0, a selector system does not need to comply with all the design/ operating parameter ranges listed in the selector diagnostic tool's results tables to control filamentous bulking. The East Bay Municipal Utility District (EBMUD) selector worked well and only complied with three parameters. Since the parameters are listed in order of their influence on diluted sludge volume index (DSVI), those listed first in the diagnostic tool's results table are those that the selector operator or designer should be primarily concerned with.

A more detailed summary of this study's findings is presented in the next section.

ES.2 Project Objectives

Selector processes have been widely applied to control filamentous bulking in activated sludge systems for more than thirty years. Still, the literature does not provide a consistent set of selector process design or operating guidelines. Variation in the degree of sludge settleability control achieved, represented by the sludge volume index (SVI), for similar process designs has dictated that selectors be designed on an empirical basis, relying heavily on design concepts and demonstrated performance at facilities with similar wastewater characteristics and process configurations.

The primary objectives of this research were to:

- Investigate the mechanisms limiting the ability of a selector to control the growth of specific filamentous organisms known to cause bulking;
- Establish a project database of selector design and performance from a large pool of full-scale facilities from across the U.S.;
- Identify selector design and performance relationships for each of the three main selector categories (aerobic, anoxic, and anaerobic) based on the project database information collected; and
- Demonstrate the implementation of a full-scale anaerobic selector at two wastewater treatment facilities and identify associated selector design and performance issues.

ES.3 Project Approach

Based on the project objectives, this study was divided into five main project tasks:

1. Literature Review
2. Laboratory Investigation
3. Initial Plant Screening Survey
4. Detailed Plant Investigations
5. Full-Scale Demonstration Projects

Dr. H. David Stensel (University of Washington, Seattle) conducted a literature search and review of selector-related topics, including filament type and occurrence in activated sludge systems, kinetic and metabolic substrate removal mechanisms, available full-scale selector design and performance data, and current research efforts related to the control of specific filamentous organisms. The goal of the literature review was to highlight key issues for application to subsequent project tasks.

The literature review illustrated that selector design approaches are focused on the removal of raCOD, while some selector designs often fail if process conditions favor the growth of filamentous organisms that thrive on slowly assimilable chemical oxygen demand (saCOD). In order to further examine this issue, under Dr. Stensel's direction, Gang Xin conducted a bench-scale, laboratory experiment to investigate the ability of an aerobic selector to control two specific filamentous organism types—one that prefers raCOD (Type 021N, *Thiothrix*) and one that thrives on saCOD (*Microthrix parvicella*, Type 0092).

This study focused primarily on collecting and analyzing selector design and operating data from full-scale facilities from across the U.S. As an initial step, a screening survey form, designed to be completed in a relatively short time period, was distributed to a large number of wastewater facilities from across the country. The initial screening survey was used to establish plant contacts at a large pool of facilities equipped with selectors of various types, collect basic selector design (type, configuration) and performance data (SVI), and identify candidate facilities interested in participating further in the study.

Following completion of the initial screening survey, many of the facilities were carried forward as part of a detailed plant investigation task. During this phase, plants were asked to provide more detailed information regarding both plant and selector design and operation, including one year of plant operating and selector performance data. Based on the activated sludge operating data provided, a number of important design parameters were calculated in order to compare selector design and performance between facilities and selector types. For the purposes of this study, the diluted SVI (DSVI) was selected as the most accurate representation of sludge settleability at these facilities because of the dependency of the SVI test on mixed liquor suspended solids (MLSS) concentration. Single variable regression analyses were conducted to evaluate the relationship between a wide array of process variables and the DSVI achieved (dependent variable). The results were compared to literature design and operating guidelines whenever possible.

Since selectors are often installed as retrofits to existing facilities rather than included in original plant designs, this study included the performance demonstration of full-scale anaerobic selectors installed at two wastewater treatment facilities—the EBMUD Main Wastewater Treatment Plant (MWWTP) in Oakland, Calif., and the Orange County Sanitation District (OCSD) Plant No. 1 in Fountain Valley, Calif. The goal of this work was to provide municipalities with key information necessary for successful selector implementation at their facilities by highlighting process considerations and issues.

ES.4 Literature Review

The following is a summary of the main literature review findings:

- A combined survey of 270 U.S. facilities (Jenkins et al. 2004) indicated that the most common filament types were (in order of frequency of occurrence) Type 1701, Type 021N, and *Thiothrix*, while a survey of 33 long-MCRT, biological nutrient removal (BNR) plants in South Africa (Blackbeard et al., 1987) found Type 0092, Type 0675, Type 0041, *M. parvicella*, and Type 0914 to be most common.
- Aerobic selectors promote kinetic conditions favoring preferential substrate uptake and sequestering by floc-formers over filamentous organisms. Anoxic selectors create a metabolic advantage for floc-formers, since most filamentous organisms are unable to denitrify (use nitrate as an electron acceptor) or have relatively low denitrification rates. Similarly, the feed-starve cycle employed in anaerobic selectors allows metabolic selection of floc-forming, phosphorus-accumulating organisms (PAOs) or glycogen-accumulating organisms (GAOs) over filamentous organisms.
- Selectors will be most successful in situations where the target filaments use raCOD as substrates. Selectors may fail if the target filament uses saCOD or sulfide or is favored by low pH or nutrient deficient conditions.
- Some filament types, such as *M. parvicella*, use saCOD [long-chain fatty acids (LCFAs)] for substrate and will proliferate in selector systems under the following conditions: zero or low dissolved oxygen (DO), long MCRT, and low temperature.
- A review of pilot- and full-scale selector design and operating data showed that a wide range of SVI control was achieved, with some installations reporting no significant improvement in bulking control. Single-stage designs are used for anoxic and anaerobic selectors, while most aerobic selectors include a staged design.

The following is a summary of general selector design guidelines found in the literature:

- Substrate Removal – The soluble COD (sCOD) leaving the selector should be <60 mg/L (Shao and Jenkins, 1989) and the raCOD should be virtually absent. The selector should remove 80% of the removable COD (Chudoba and Wanner, 1987).
- Selector Staging and Configuration – All three selector types (aerobic, anoxic, anaerobic) should be designed with at least three stages, sized at 25%, 25%, and 50% of the total selector volume, respectively (Jenkins et al., 2004). A staged-selector arrangement is necessary to create a food-to-microorganism (F/M) gradient (Albertson, 2005).
- Aerobic Selectors – Aerobic selectors should be staged to provide proper kinetic conditions favoring rapid substrate uptake and storage by floc-formers over filaments. Jenkins et al. (2004) recommended a three-stage design, sized at 25%, 25%, and 50% of the total selector volume with first stage and total F/M loadings of 12 kg COD/(kg MLSS·d) and 3 kg COD/(kg MLSS·d), respectively.
- Anoxic Selectors – In single-stage arrangements, the selector F/M should be $\leq$1 kg BOD_5/(kg MLSS·d) for temperatures $\leq$18°C and $\leq$1.5 kg BOD_5/(kg MLSS·d) for temperatures >18°C, while the anoxic MCRT should be at 1-2 d (Marten and Daigger, 1997). Grady et al. (1999) recommended an anoxic MCRT of 1.0 d at temperatures >20°C and 1.5 d at temperatures <17°C. Jenkins et al. (2004) recommended a three-stage design, sized at 25%, 25%, and 50% of the total selector volume with first-stage and total F/M loadings of 6 kg COD/(kg MLSS·d) and 1.5 kg COD/(kg MLSS·d), respectively.
- Anaerobic Selectors – A three-stage selector with a total selector hydraulic residence time (HRT) of 0.75–2.0 h is recommended (Jenkins et al., 2004).

ES.5 Laboratory Investigation

Four 3-L bench-scale, completely mixed activated sludge (CMAS) units (R1, R2, R3, and R4) were initially seeded with activated sludge containing both *Thiothrix* spp. (raCOD filament) and *M. parvicella* (saCOD filament). The reactors were fed a synthetic wastewater high in Tween 80 (water soluble oleic acid ester of sorbitol) and acetate to promote the growth of both raCOD and saCOD filament types. After an initial startup period, the following changes were made: 1) a three-stage aerobic selector was added to R1 (25%, 25%, and 50% of total selector volume), 2) the raCOD constitutents were removed from the feed to R2, and 3) a four-stage aerobic selector was added to R4 (12.5%, 12.5%, 25%, and 50% of total selector volume). No changes were made to R3, which served as the control. Oxygen uptake rate (OUR) batch tests were conducted periodically by adding either acetate (raCOD) or Tween 80 (saCOD) to mixed liquor samples from each reactor. The reactor operating conditions are summarized in Table ES-1.

Table ES-1. Summary of Bench-Scale Reactor Operating Conditions.

Reactor No.	Description	Operating Conditions (all reactors) Wastewater Feed	MCRT(d)	Temp. (°C)	Air Feed
1	Three-stage aerobic selector	Synthetic, high in LCFAs (oleic acid) and raCOD (acetate)	20	12–15	Intermittent, DO between 0–2 mg/L
2	raCOD removal from feed				
3	Single-stage CSTR				
4	Four-stage aerobic selector				

The DSVI variation over time in each of the four bench-scale units is shown in Figure ES-1. The results suggest that adding a three-stage and four-stage aerobic selector to R1 and R4, respectively, had a similar effect on DSVI reduction as removing raCOD from the feed to R2. The systems equipped with selectors, however, actually achieved slightly improved DSVI values, suggesting that aerobic selectors may do more to control bulking than just remove raCOD. Severe bulking occurred in the control reactor with *Thiothrix* spp. as the dominant filament type. Conditions favoring the growth of *M. parvicella* could not be maintained in any of the reactors.

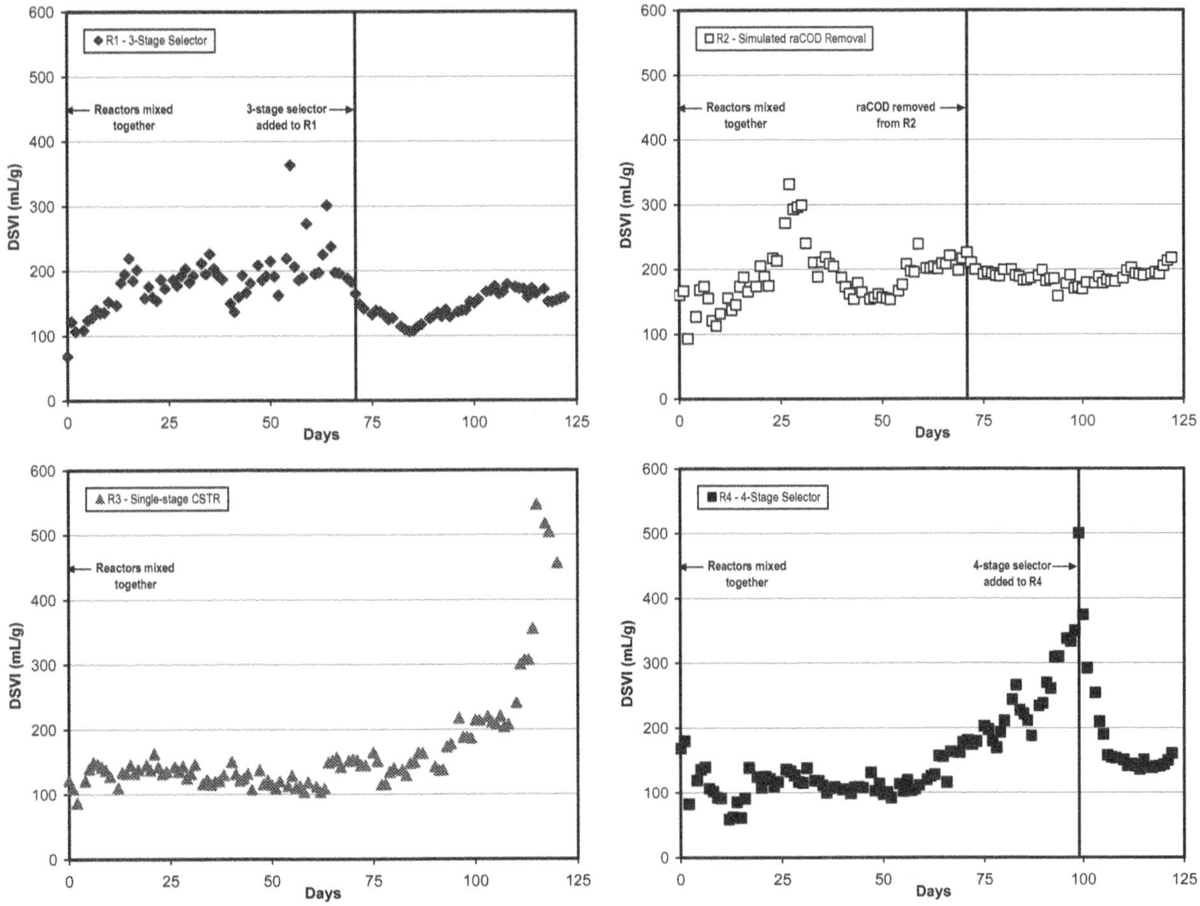

Figure ES-1. Dilute Sludge Volume Index in Four Bench-scale Reactor Systems.

OUR and acetate uptake rates were dramatically reduced in R2 following raCOD removal from the wastewater feed and were significantly less (2-6 times lower for OUR, 3-7 times lower for acetate) than the other reactors. This suggests that the R2 feed without raCOD did not support raCOD floc-forming bacteria growth and that the presence of these bacteria may enhance floc structure and settleability. Acetate uptake rates were 6–10 times higher than the Tween 80 uptake rates,which suggests that Tween 80 (and possibly all LCFAs) may not be adequately removed in a selector and could leak into the main aeration zone at sufficient levels to support filamentous bulking. Similar DSVI control was achieved in both the three- and four-stage aerobic selector systems, while sCOD profiles indicated that most of the removal occurred in the first stage.

ES.6 Initial Plant Screening Survey

The initial screening survey included 125 U.S. wastewater treatment plants. Of these facilities, 85 had selectors (aerobic, anoxic, or anaerobic), but only 46 had improved settleability following selector installation, as shown in Figure ES-2.

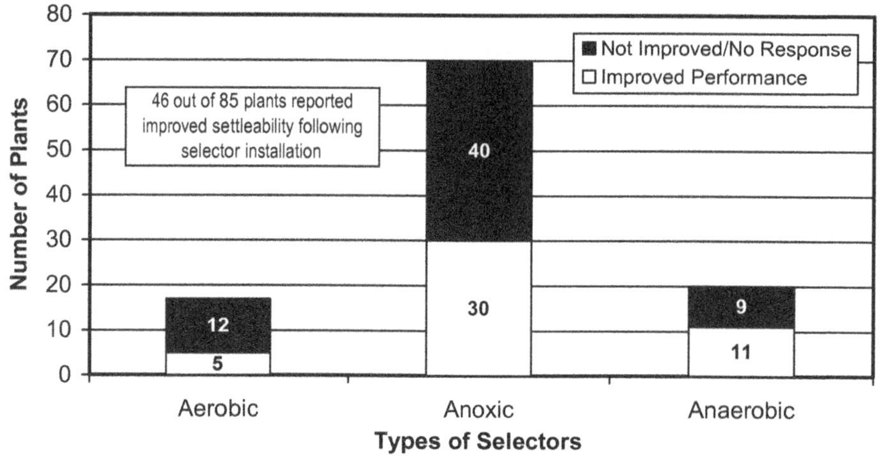

Figure ES-2. Initial Screening Survey Results – Selector Type and Effectiveness.

The initial screening survey form requested the following basic plant information:

- Plant flow rate
- MCRT
- Nutrient removal requirements
- Aeration basin configuration
- Type of selector
- Bulking frequency
- SVI control achieved following selector installation

Given the significant amount of additional plant data to be requested and assuming a moderate response rate, the project team decided to carry forward all 85 facilities reporting selector installations to the detailed plant investigation phase.

ES.7 Detailed Plant Investigations

Table ES-2 summarizes the information requested from each of the 85 facilities included in the detailed plant investigation. In addition to collecting general plant and process configuration information, each facility was asked to provide approximately one year of selector operating and performance data in spreadsheet format. The extensive data collection effort required numerous follow-up data requests and discussions with plant contacts to verify the information provided and to answer plant-specific questions. A number of important selector design and operating parameters were calculated based on the information provided by each plant, as summarized in Table ES-3.

Most facilities reported sludge settleability performance on an SVI basis. Given the dependence of the SVI test result on mixed liquor concentration, as reported by Dick and Vesilind (1969), reported SVI values were converted to DSVIs by applying a correction developed by Merkel (1971). Lee et al. (1983) reported that the DSVI test yielded the best correlation with total extended filament length relative to other techniques for estimating sludge settleability.

Table ES-2. Summary of Detailed Plant Investigation Data Requested.

Category	Description
General Information	• Facility name, location, contact • Average, peak flow rate • Industrial contribution, major contributors • Annual wastewater temperature range • Nutrient removal requirements and processes
Selector Configuration	• Selector type (aerobic, anoxic, anaerobic) • Number and volume of selector stages • Mixing type (hydraulic, mechanical, air) • Available process design criteria, technical reports
Aeration Basin Configuration	• Number and volume of aeration stages and basins • Type of aeration system • Internal recycle streams • Approximate DO profiles • Location of RAS feed points
Additional Plant Information	• Process schematic • Secondary process operation and maintenance (O&M) manuals • Secondary influent sulfide levels • Oxygen uptake rate data • Soluble BOD or COD exiting the selector zone
Plant Operating Data (One Year)	• Secondary influent – flow, BOD, sBOD, COD, sCOD, TKN, P • Number of aeration basins in-service • WAS, RAS flow and concentration • MLSS, MLVSS • System (excluding clarifier solids), aerated MCRT • F/M • DO • Influent or effluent pH • SVI or DSVI • Filament type and abundance • RAS chlorination periods

Table ES-3. Summary of Detailed Plant Investigation Process Data Calculations.

Parameter	Comments
Selector MCRT (d)	Calculation based on mass of mixed liquor in selector zone only
Contact (or floc) loading (kg BOD_5/kg MLSS)	Ratio of influent BOD mass to solids mass in initial contact zone (ICZ)
Selector ICZ F/M loading [kg BOD5/(kg MLSS·d)]	F/M calculation based on mass of mixed liquor in selector ICZ only
Selector HRT (h)	HRT calculation based on volume of selector zone only
90th Percentile SVI (mL/g)	
90th Percentile Merkel DSVI (mL/g)	SVI data converted to DSVI using Merkel equation
Fraction of SVIs greater than 150 mL/g (%)	Represents percent of time SVIs exceed typical control limit

Given the large amount of information requested from each facility, many facilities were not able to provide key information, such as filament type and abundance, SVI, or essential secondary process operating data. Despite this limitation, the study was successful in collecting and verifying data from 44 of the 85 original plants for a total of 48 data sets (four facilities included two data sets representing distinct operating modes). The facility size and selector type distribution is presented in Figure ES-3. A tabular summary of all data collected is included in Table 4-5 in the main report.

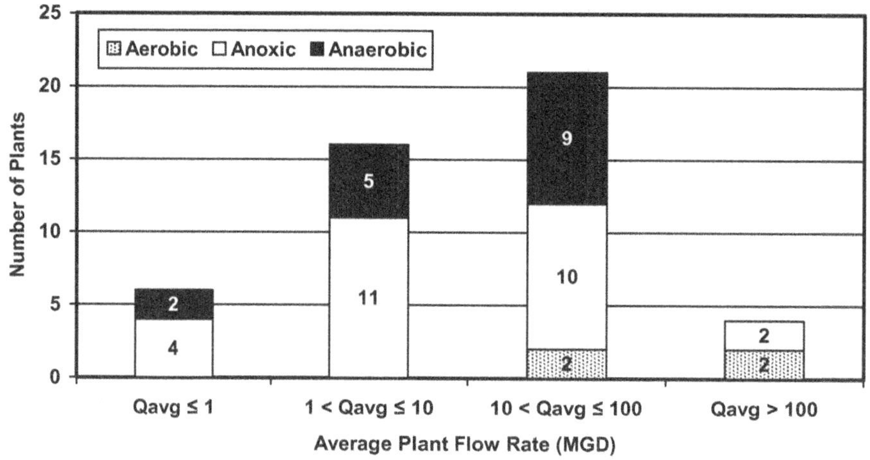

Figure ES-3. Facility Size, Selector Type Distribution.

Average values for a number of selector process design parameters were plotted against both 90th percentile SVI and DSVI results (including the Merkel correction, as necessary). Figure ES-4 is a plot of the ICZ F/M, selector F/M, selector MCRT, system MCRT (excluding clarifier solids), total selector HRT, and number of selector stages versus 90th percentile DSVI. For the purposes of this study, a 90th percentile DSVI value of 150 mL/g was selected as the typical upper limit for well-settling sludge.

The plots in Figure ES-4 clearly indicate that the anoxic selectors achieved greater bulking control relative to the anaerobic selectors. Nearly all of the anoxic selector facilities (23 of 27) had 90th percentile DSVIs <150 mL/g, while nearly all of the anaerobic selector plants (12 of 14) exceeded this limit. Two of five aerobic selector plants also exceeded 150 mL/g. Most anoxic selectors, however, were installed in long-MCRT plants, while all anaerobic selectors were installed in short-MCRT plants (see Figure ES-4). Therefore, the lower DSVI in plants with anoxic selectors may be because of the lower DSVI produced by long-MCRT filamentous bacteria (Wanner, 1994), rather than selector type.

No clear relationships were observed between settleability control and selector ICZ F/M, selector F/M, selector MCRT, system MCRT (excluding clarifier solids), or total selector HRT. In fact, a wide range of DSVIs was observed across a broad range of F/M loading rates, MCRTs, and selector HRTs. Selector staging was not observed to have a significant impact on bulking control in the anoxic selector systems. All eight single-stage anoxic selectors yielded DSVIs <150 mL/g, while four of 18 multi-stage anoxic selectors exceeded this limit. Selector staging was also not observed to have a significant impact on settleability in anaerobic selector systems, since six of seven plants yielded DSVIs >150 mL/g in both the single- and multi-stage categories.

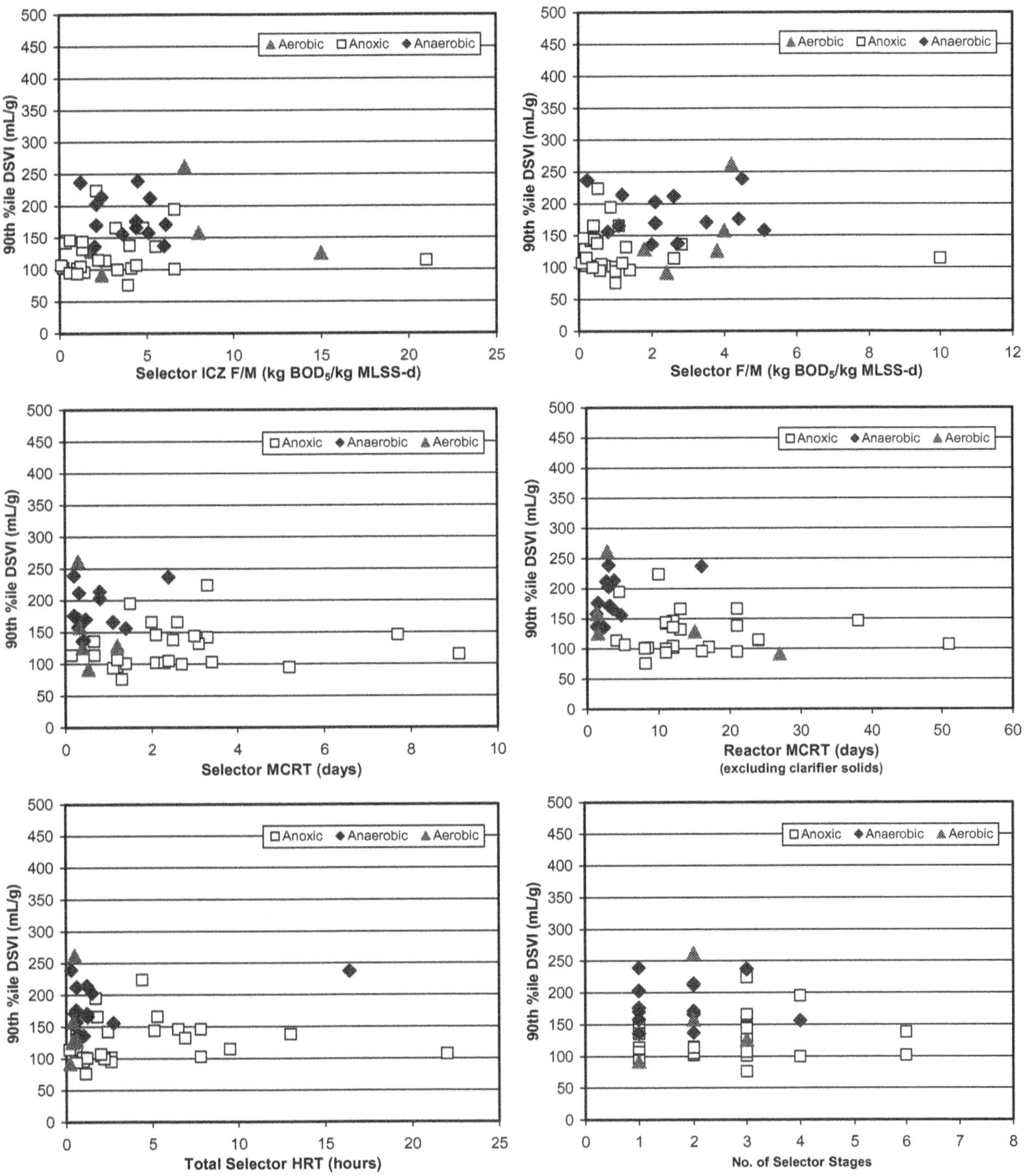

Figure ES-4. Average Selector ICZ F/M, Selector F/M, Selector MCRT, System MCRT, Total Selector HRT, Number of Selector Stages versus 90th Percentile DSVI.

Comparing average parameter and 90th percentile SVI/DSVI values for the plants included in the detailed plant investigation is limited since each facility is represented by only a single data point and does not reflect variation in each parameter. A single-variable regression analysis, incorporating daily operating data for each facility, was conducted to better evaluate the influence of parameter variation on SVI and DSVI values.

Regression Analysis

The 44 full-scale facilities participating in the detailed plant investigations were divided into three distinct plant categories—short-MCRT with anoxic or anaerobic selectors, short-MCRT with aerobic selectors, and long-MCRT. Short- and long-MCRT plants were separated because it has been shown that dominant filamentous bacteria in short-MCRT systems have significantly different growth requirements than those in long-MCRT systems. The literature also suggests that selectors are not as effective in controlling long-MCRT filaments (Gabb, 1988; Gabb *et.al.*, 1991; Wanner, 1994; Jenkins et al., 2004; Martins et al., 2004b). Because aerobic selectors primarily rely on kinetic mechanisms, while anoxic and anaerobic systems may use a combination of metabolic and kinetic mechanisms, short-MCRT WWTPs with aerobic selectors were separated from short-MCRT WWTPs with unaerated (anoxic or anaerobic) selectors in the analysis.

Single-variable regression analyses were conducted regressing a wide array of process parameters (independent variable) against DSVI (dependent variable). A ranking of the most significant process variables for each plant category was developed based on regression analysis R^2 values, which is the percent variation in the dependent variable accounted for by variation in the independent variable. Cubic polynomial regression curves were calculated and shown on regression plots for each of the variables to identify ranges that may influence DSVIs in selector-equipped facilities. Using the regression analysis, the project team developed recommendations for the significant process variables and compared these to literature values. A summary of the recommended parameter ranges for the short-MCRT WWTPs with anoxic or anaerobic selectors, short-MCRT WWTPs with aerobic selectors, and long-MCRT plants is presented in Tables ES-4, ES-5, and ES-6, respectively. The parameters in each table are listed in order of strongest influence on DSVI (R^2 value).

Based on the analysis of short-MCRT WWTPs with anoxic and anaerobic selectors, selectors in this category should be sized large enough to remove all or most of the raCOD and should be staged to prevent short-circuiting and raCOD breakthrough to the main aeration basin but not to provide a kinetic advantage. Increasing the selector ICZ F/M was shown to increase DSVI, while increasing the number of selector stages resulted in a reduction in DSVI.

Table ES-4. Recommended Parameter Ranges for Short-MCRT WWTPs with Anoxic or Anaerobic Selectors.

Parameter	Recommendations from this Study	Recommendations from Literature	Literature References
Average MLSS (mg/L)	1,500-2,000+		
Reactor MCRT (d)	>4.5		
Total Selector F/M [kg BOD_5/(kg MLSS·d)]	<1.0 (lower the better)	≤1.0	Jenkins, 2004
Selector MCRT (d)	2-3+	1.0-2.0	Jenkins, 2004
Number of Selector Stages	2	3	Jenkins, 2004; Wanner, 1994
Aeration Basin DO (mg/L)	2.5-4.0 (air plants only)	>1-2	Jenkins, 2004; Wanner, 1994
BOD/TSS Ratio[1]	<0.5 (lower is better)		
ICZ F/M [kg BOD_5/(kg MLSS·d)]	<1.0 (lower the better)	~3	Jenkins, 2004
Selector HRT (without recycle) (h)	min. of 1.2, >2.5 best		
Selector Vol/Total Basin Vol Ratio (%)	22.5-25.0	25	Wanner, 1994
Selector HRT (with recycle) (h)	>1.5	0.75-2.0	Jenkins, 2004
ICZ HRT w/RAS (h)	1.4-1.6		
ICZ HRT w/o RAS (h)	2.4-2.7		
Effluent Temperature (°C) [1]	20-25 (27-30+ worst)		
Number of Aeration Basin Stages	not significant		
Act Sldg. Inf. BOD (mg/L)	not significant		
%RAS Flow (%)	not significant	≤100	Wanner, 1994
Effluent pH	not significant		

Note: [1] Best results found in this range, but making adjustments to operate in this range is not recommended.

For short-MCRT WWTPs with aerobic selectors, DSVI decreases with decreasing ICZ HRT and increases with increasing influent BOD_5 concentration. This analysis supports the hypothesis that a concentration gradient is required to provide a kinetic advantage to floc-formers over filamentous organisms; however, at higher influent BOD_5 concentrations, sufficient raCOD may leak through to the main aeration zone to cause bulking problems.

Table ES-5. Recommended Parameter Ranges for Short-MCRT WWTPs with Aerobic Selectors.

Parameter	Recommendations from this Study	Recommendations from Literature	Literature References
Act. Sldg. Inf. BOD (mg/L)	<80	N/A	
ICZ HRT (without recycle) (min)	4.5-7.5		
ICZ HRT (with recycle) (min)	3.5-6.0		
Effluent pH	6.3-6.6	N/A	
Total Selector HRT (without recycle) (min)	≤18		
% RAS Flow (%)	25-35	≤100	Wanner, 1994
Total Selector HRT w/RAS (min)	15-18	10-20	Wanner, 1994
Effluent Temperature (°C)	<18-19 (worst: 21-23+)	<28	Wanner, 1994
ICZ F/M [kg BOD_5/(kg MLSS·d)]	~15	~5-6 ≥16 ok	Jenkins et al., 2004 Wanner, 1994
Reactor MCRT (d)	<1.3		
Average MLSS (mg/L)	max. of 1,000		
Aeration Basin DO (mg/L)	14-18 (pure O_2 plants)	>10 (pure O_2 plants)	Wanner, 1994
Total Selector F/M [kg BOD_5/(kg MLSS·d)]	not significant	~1.5-2.0	Jenkins et al., 2004
Number of Selector Compartments	N/A[1]	3	Wanner, 1994; Jenkins, 2004

[1] Insufficient data variation in data set to adequately assess this parameter.

Based on the regression analysis, anoxic selectors do not appear to significantly control filamentous organisms and bulking in long-MCRT plants, which is supported by the literature (Wanner, 1993; Jenkins et al., 2004; Martins et al., 2004b). The analysis suggests that DSVIs will be controlled best by using a selector HRT, ICZ HRT, and number of stages approaching zero (→ 0), which means that removing the anoxic selector (or anoxic zones) is best for reducing DSVI in long-MCRT systems.

Table ES-6. Recommended Parameter Ranges for Long-MCRT Plants.

Parameter	Recommendations from this Study	Recommendations from Literature	Literature References
Average MLSS (mg/L)	2,500-4,500+		
Selector HRT (with recycle) (h)	→ 0	0.75-2.0	Jenkins et al., 2004
ICZ HRT (without recycle) (min)	→ 0		
Selector Vol/Total Basin Vol.Ratio	→ 0	0.25	Wanner, 1994
Number of Aeration Basin Stages	more is better, up to 8	many	Jenkins et al., 2004
Selector HRT (without recycle) (h)	→ 0		
Number of Selector Stages	0	3	Jenkins et al., 2004; Wanner, 1994
Effluent pH	6.4-6.7 best (7.7+ worst)		
Effluent Temperature (°C)	27-32 best (13-17 worst)		
% RAS Flow (%)	not significant	≤100	Wanner, 1994
Activated Sludge Influent BOD (mg/L)	not significant		
Aeration Basin DO (mg/L)	not significant	>1-2	Jenkins et al., 2004; Wanner, 1994
ICZ HRT (without recycle) (h)	not significant		
Reactor MCRT (d)	not significant		
Selector F/M [kg BOD_5/(kg MLSS·d)]	not significant	≤1.0	Jenkins et al., 2004
BOD/TSS Ratio	not significant		
Selector MCRT (d)	not significant	1-2	Jenkins et al., 2004
ICZ F/M [kg BOD_5/(kg MLSS·d)]	not significant	6	Wanner, 1994

The information provided in Tables ES-4, ES-5, and ES-6 has been incorporated into a computerized selector diagnostic tool for assistance in troubleshooting existing selector installations or designing new selectors based on user input and design and operating parameter recommendations from this study and the literature. A CD-ROM containing this software can be found on the inside back cover of this report. Instructions for using the software application, including screen shots, is included in Appendix F.

ES.8 Full-Scale Demonstration Projects

Full-scale, pilot anaerobic selector evaluations were conducted at the EBMUD MWWTP and OCSD Plant No. 1 from June to October 2003 and July to November 2004, respectively. Both plants were operated in split-plant mode with a selector-equipped plant and a control plant throughout each test period. A summary of the selector operating and performance data at each facility is presented in Table ES-7.

Table ES-7. Comparison of EBMUD and OCSD Selector Operating and Performance Data.

Parameter	EBMUD	OCSD[1]	Recommendations from this Study	Literature Value	Literature Reference
Plant Type	high-purity oxygen	air			
Selector Type	anaerobic	anaerobic			
Flow (MGD)	34.3	24.5			
SVI (mL/g)					
Average	120	345			
90th Percentile	166	536			
Orthophosphate (mg-P/L)					
Secondary Influent	4.1	5.1			
Stage 1 (anaerobic)	12.0	9.1			
Dominant Filament Types	Type 021N	Type 021N, *Thiothrix*, Type 1701, *S. natans*			
MLSS (mg/L)	2,040	831	1,500–2,000+		
Reactor MCRT (d)[2]	1.3	1.9	High as possible		
Selector F/M [BOD_5/(kg MLSS·d)]	5.1[3]	4.3[4]	<1.0 (lower is better)	≤1.0	Marten and Daigger (1997)
Selector MCRT (d)	0.32	0.32	2–3+	1–2	Marten and Daigger (1997)
Number of Selector Stages	1	1	2	3	Jenkins, 2004; Wanner, 1994
Aeration Basin DO (mg/L)	N/A	1.4	2.5–4.0 (air plants only)	>1-2	Jenkins et al. (2004)
Sec. Inf. BOD_5/TSS Ratio	2.4[3]	2.4[4]	<0.5		
Avg. Selector HRT (w/RAS/w/o RAS) (min)	26/34	38/67	>90/>150	45–120 (w/o recycle)	Jenkins et al. (2004)
Selector Volume to Total Basin Volume Ratio (%)	25	17	22.5-25.0	25	Wanner, 1994
Temperature (°C)	27	27	20-25 (27-30+ worst)		
Total System F/M [BOD_5/(kg MLSS·d)]	1.3[3]	0.7[4]	not significant		

Notes: [1] Based on Phase 4 data.
[2] Excludes secondary clarifier solids.
[3] Value reported is on a $cBOD_5$ basis and converted to BOD_5 using $BOD_5 = 1.45 \times cBOD_5$.
[4] Value reported is on a COD basis and converted to BOD_5 using $BOD_5 = 0.5 \times COD$.

The EBMUD MWWTP selector plant produced an average and 90th percentile SVI of 120 and 166 mL/g, respectively, which compared favorably to the respective control plant SVI values of 270 and 471 mL/g. Conversely, the OCSD Plant No. 1 anaerobic selector system yielded significantly higher SVIs compared to the control plant, with no single reported SVI value <200 mL/g. Although the EBMUD selector was operated at the same MCRT as the OCSD installation, significantly more orthophosphate release and lower DSVIs occurred with the EBMUD selector. OCSD reported a number of possible factors that may have influenced the poor SVI control, including problems with air feed to the selector plant and the presence of sulfur granules in both Type 021N and *Thiothrix* (possible secondary influent septicity). As shown in Table ES-7, the OCSD aeration basin average DO level (1.4 mg/L) was significantly lower than the range recommended by this study (2.5-4.0 mg/L), while EBMUD's oxygen vent gas purities were very high (70-80^{+}% throughout the study).

Based on the data presented in Table ES-7, the following conclusions were made regarding the reported differences in SVI control:

- Selector HRT does not appear to be a significant factor, since the EBMUD selector HRT was actually shorter than the OCSD selector (35 min vs. 45 min).
- Because both systems were operated at roughly the same selector MCRT, this does not appear to be a significant factor.
- The OCSD selector was operated at a significantly lower MLSS concentration (831 mg/L), which may have contributed to the reduced phosphorus release and uptake relative to the EBMUD selector (MLSS = 2,040 mg/L). This study recommends a MLSS value of 1,500–2,000+ mg/L for improved settleability.
- The EBMUD and OCSD selectors were sized at 25% and 17% of the total reactor volume, respectively. The regression analysis results predict that the EBMUD selector would yield improved settleability relative to the OCSD selector installation due to the increased ratio of selector to total reactor volume.
- A selector can be successful even though it is not operated within all the important design/operating parameters ranges as identified by this study and the literature; however, if a selector is operated outside of all the recommended parameter ranges, it is unlikely that the selector will be successful.

ES.9 Summary and Conclusions

Although the literature provides separate design and operating parameters for aerobic, anoxic, and anaerobic selectors, these parameters are assumed to be the same for either long- or short-MCRT activated sludge plants. Since distinctly different groups of filamentous bacteria predominate at short- versus long-MCRT—due in large part to differences in growth requirements between these two filament groups—they may require different control parameters. Consequently, the full-scale activated sludge plant data collected during this study were separated into long- and short-MCRT groups, based primarily on the type of filamentous bacteria present.

The short-MCRT plants were further split into two groups—plants equipped with aerobic selectors and plants equipped with either anoxic or anaerobic selectors—based on the hypothesis that aerobic selectors were more kinetically favorable to floc-forming bacteria than filamentous bacteria, and anaerobic or anoxic selectors were more metabolically favorable to floc-forming bacteria. The results of the regression analysis supported these differences among the three different WWTP groups—short-MCRT with anoxic or anaerobic selectors, short-MCRT with aerobic selectors, and long-MCRT. In general, selectors in the long-MCRT plants did not appear to reduce filamentous bulking (DSVI); in fact, the results suggest that unaerated selectors may enhance filamentous bulking in long-MCRT plants.

Selector design and operating parameters were quantitatively ranked according to their influence on DSVI using the regression R^2 value. Using cubic polynomial regression curves, parameter ranges that were associated with the lowest (and highest) DSVI were determined. Many of these parameter ranges agreed well with those found in the literature. Some did not, for reasons that could be explained logically. Some design and operating parameters thought to be significant at the start of this study were instead found to have little if any influence on DSVI.

The regression analysis suggests that for the lowest DSVI, the best long-MCRT designed and operated plants had high MLSS (2,500-4,500$^+$ mg/L), compartmentalized aeration basins, and no anoxic or anaerobic zones (if nutrient removal was not needed). Further, DSVIs were lower when pH = 6.4–6.7 and temperatures = 27-32°C, and higher when pH = 7.7$^+$ and temperature = 13-17°C.

According to the regression analysis (see Table 4-10), the best design and operation of a short-MCRT activated sludge plant with anoxic or anaerobic selectors would include: a selector volume as large as possible while keeping the selector volume to total basin volume ratio between 22.5-25.0%, two selector stages, a selector MCRT >2–3$^+$ d, a MLSS concentration of 1,500–2,000$^+$ mg/L, an aeration basin DO between 2.5 and 4.0 mg/L, and as long a reactor MCRT as possible. Other factors influencing DSVI include activated sludge influent BOD_5/TSS ratio (best is <0.5), and effluent temperature (best is 20-25°C and worst is 27-30°C, which matches well with the Type 1701 growth rate being higher than floc-forming bacteria at temperatures around 28°C and frequently less than floc-formers at temperatures less than 28°C—per Wanner, 1994).

The aerobic selector ICZ must be small enough in short-MCRT plants to provide a high enough raCOD to induce kinetic selection of floc-forming bacteria over filamentous bacteria. Although higher influent BOD_5 concentrations may result in raCOD bleeding through a selector, the ICZ F/M does not appear to be the most important design and operating parameter for a successful aerobic selector. Further, the %RAS should be as low as possible (25-35%), the MLSS should be as low as possible (to about 1,000 mg/L), the reactor MCRT should be low (<1.3 d), and the aeration basin DO should be high.

The following is a summary of additional conclusions from this study:

Laboratory Investigation

- Severe bulking (DSVI ≥500 mL/g) due to *Thiothrix* spp. was controlled by installing three-stage and four-stage aerobic selectors and by removing raCOD from the wastewater fed to an activated sludge process. This suggests that removing raCOD prior to the main activated sludge aeration basin, either with a selector or by excluding it from a synthetic sewage fed to the activated sludge process, significantly reduces the growth of *Thiothrix* spp.
- Removing raCOD from wastewater fed to activated sludge processes alone may not produce DSVIs as low as activated sludge processes equipped with a well-performing selector. This may be because selectors enhance the growth of raCOD floc-forming bacteria, while activated sludge processes fed wastewaters absent of raCOD do not support the growth of these organisms; and raCOD floc-forming bacteria may enhance activated sludge floc structure and settling on their own.
- Uptake rates for Tween 80, and possibly LCFAs, were 6-10 times slower than uptake

rates for acetate. This suggests that even well-performing selectors may not adequately remove LCFAs and could allow them to leak into the main aeration basin where they may be used by filamentous organisms for growth.

Detailed Plant Investigations

Comparing average selector design/operating data vs. 90th percentile SVI and DSVI yielded the following conclusions:

- Anoxic selector installations demonstrated superior settleability control compared to anaerobic selectors. Approximately 85% of anoxic selector plants (23 of 27) had 90th percentile DSVIs <150 mL/g, while only 14% of anaerobic plants (two of 14) achieved this result. Most anoxic selectors, however, were installed in long-MCRT plants, while all anaerobic selectors were installed in short-MCRT plants. The lower DSVI in plants with anoxic selectors may be because of MCRT and the type of filamentous bacteria that grow at long MCRT (Wanner, 1994) rather than selector type.
- Selector staging was not observed to have a significant impact on settleability in anoxic selector systems. In fact, all eight single-stage anoxic selectors yielded 90th percentile DSVIs <150 mL/g, while four of 18 multi-stage anoxic selectors exceeded this value.
- Selector staging was not observed to have a significant impact on settleability in anaerobic selector systems, since six of seven systems yielded 90th percentile DSVIs >150 mL/g in both the single- and multi-stage categories.
- Based on plots of average values vs. 90th percentile DSVIs, no significant relationships were identified between settleability control and selector ICZ F/M, selector F/M, selector MCRT, system MCRT (excluding clarifier solids), contact loading, or selector HRT.

Comparing average parameter and 90th percentile SVI/DSVI values for the plants included in the detailed plant investigation is limited since each facility is represented by only a single data point and does not reflect variation in each parameter. A single-variable regression analysis, incorporating daily operating data for each facility, was conducted to better evaluate the influence of parameter variation on SVI and DSVI values.

The regression analysis for short-MCRT WWTPs with anoxic or anaerobic selectors, short-MCRT WWTPs with aerobic selectors, and long-MCRT WWTPs yielded the following main conclusions:

- For short-MCRT WWTPs, anaerobic and anoxic selectors should be sized large enough to remove all or most of the raCOD and should be staged to prevent short-circuiting and raCOD breakthrough to the main aeration basin (rather than to provide a kinetic advantage).
- For short-MCRT WWTPs with aerobic selectors, a raCOD concentration gradient is required to provide a kinetic advantage to floc-formers over filamentous organisms; however, at higher influent BOD_5 concentrations, sufficient raCOD may leak through to the main aeration zone to cause bulking problems.
- Selectors do not significantly control filamentous organisms and bulking in long-MCRT plants, which is supported in the literature (Wanner, 1993; Jenkins et al., 2004; Martins et al., 2004b).

Full-Scale Demonstration Projects

Based on the comparison between the EBMUD and OCSD selector systems' operating values and recommended parameter ranges determined in this study's regression analysis, the following conclusions were made:

- Selector installations can be successful even if operated outside of some or most of the recommended design/operating parameter ranges for successful selector operation.
- Selector installations will probably not be successful if operated outside of all the recommended design/operating parameter ranges for successful selector operation.
- Average MLSS appears to be an important parameter to keep within the recommended operating range. In the EBMUD/OCSD case, the selector volume/total basin volume ratio and aeration basin DO concentration also appeared to be important parameters.
- Using the recommended parameter ranges for successful selector operation as a guide appears to offer good assistance to those who wish to determine why a selector is not performing as expected, or to optimize a selector design.

Computerized Selector Diagnostic Tool

The computerized selector diagnostic tool prepared for this project, and accessible through the CD-ROM attached to the inside back cover of this report, is an easy way to use this method for assistance in troubleshooting or designing a selector installation. Documentation for this software can be found in Appendix F of this report.

CHAPTER 1.0

INTRODUCTION

1.1 Project Background

Successful operation of the activated sludge process depends on the settleability of the mixed liquor and the ability to separate it into RAS and a high quality final effluent. Solids settleability is dictated in large part by the relative populations of floc-forming and filamentous microorganisms present in the activated sludge. During successful operation, mixed liquor will be composed primarily of floc-formers with relatively low levels of filamentous organisms. Some population of filamentous organisms can help promote the formation of larger flocs and prevent development of a "pin-floc" condition, which often results in high effluent turbidities (Jenkins et al., 2004). However, large, strong flocs with excellent settling properties can form without filamentous microorganisms being present. When process conditions favor the growth of filamentous organisms over floc-formers, filamentous "bulking" may occur. Filamentous bulking may develop in response to specific wastewater characteristics and secondary process operating and environmental conditions. Some facilities may consistently experience bulking due to a process design and configuration that favors the growth of filamentous organisms. Filamentous bulking can severely interfere with proper settling and compaction characteristics in the secondary clarifier, leading to high sludge blanket levels, low RAS concentrations, and loss of suspended solids in the secondary effluent. Sludge that settles poorly also requires higher aeration basin and secondary clarifier capacities (Parker et al., 2003), higher RAS flow rates, higher WAS and chemical conditioning aid flow rates to downstream thickening processes, and higher flows to, and recycle flows from, solids handling and treatment processes.

Operating an activated sludge aeration basin in a plug-flow regime, rather than a completely mixed, continuously fed flow regime, has been demonstrated in many studies to control filamentous bulking in activated sludge systems (Chambers, 1982; Grau et al., 1982; Lee et al., 1982; Rensink et al., 1982). Installing one or more small mixing basins, called selectors, for RAS and influent wastewater prior to the main aeration basin has been shown to provide sufficient plug-flow characteristics to inhibit the growth of filamentous organisms and control bulking in activated sludge systems (Chudoba et al., 1973a; van Niekerk, 1985). Selectors have not been universally successful for controlling filamentous organisms in activated sludge systems (Osborn et al., 1986; Wakefield and Slim, 1987; Gabb, 1988; Daigger and Nicholson, 1990; Gabb et al., 1991).

Much of the early work supporting the effectiveness of selectors was performed with laboratory-scale activated sludge units. These laboratory units predominantly grew the filamentous organisms *Thiothrix* spp., Type 021N, *Sphaerotilus natans*, Type 1701, and sometimes Type 0961, when operated in the completely mixed mode. Gabb et al. (1988) suggested that *S. natans* may proliferate in laboratory units only because of the greater surface area to volume ratios of laboratory-scale units and because of the large surface area present in feed lines. Under the same operating parameters, these filamentous organisms may not grow in full-scale activated sludge plants.

Thiothrix spp., Type 021N, S. *natans*, Type 1701, and Type 0961 appear to thrive predominantly on readily assimilable chemical oxygen demand (raCOD –see discussion starting at the bottom of page 1-3) (Eikelboom, 1977; Hao 1982; Richard et al., 1982; Richard et al., 1983; Richard et al., 1984). Lee et al. (1982) reported that the growth of filamentous bacteria in laboratory-scale, completely mixed activated sludge (CMAS) units did not occur when fed a relatively weak settled domestic wastewater. When this wastewater was mixed with raw domestic sludge, settled, and then fed to the CMAS system, filamentous bulking occurred. van Niekerk (1985) used this method of supplementation to ensure filamentous bulking and suggested that supplementation with raw sludge enriched the settled domestic wastewater with soluble, short chain organic acids. He achieved the same filamentous organism growth (predominantly Type 021N) in laboratory-scale activated sludge units by supplementing the same domestic wastewater with sodium acetate.

Many researchers have used substrates high in raCOD rather than domestic wastewater to produce filamentous bulking activated sludge in CMAS systems (Chudoba et al., 1973b; Houtmeyers et al., 1980; Verachtert et al., 1980; Chiesa et al., 1982; Lee et al., 1982; Rensink et al., 1982; van den Eynde et al., 1982; Wu et al., 1983; van Niekerk, 1985). Verachtert et al. (1980) could not grow filamentous organisms in a CMAS system on a synthetic wastewater containing casein as its major carbonaceous substrate, and no raCOD.

Slijkhuis (1983) showed that *Microthrix parvicella* grew predominantly on slowly assimilable COD (saCOD), such as long-chain fatty acids (LCFAs), and poorly or not at all on raCOD, such as glucose and acetate. Gabb et al. (1991) showed that *M. parvicella* is not controlled by a properly operating aerobic selector with the following characteristics:

- Higher batch initial oxygen consumption rate and soluble COD (sCOD) uptake rate compared to an activated sludge grown in a CMAS system (Chudoba et al., 1973b; Houtmeyers et al., 1980; Verachtert et al., 1980; van den Eynde et al., 1982; Jenkins et al., 1983; Daigger et al., 1983; Wheeler et al., 1983; van Niekerk, 1985; Still et al., 1986);
- Removal of all or most of the raCOD in the selectors prior to the main aeration basin (van Niekerk, 1985; Shao, 1986); and
- Significant levels of zoogloeal colonies (van Niekerk, 1985).

Three general types of selectors have been defined—aerobic, anoxic, and anaerobic. Each provides an environment where floc-forming bacteria gain an advantage over filamentous organisms through their ability to competitively take up and store raCOD. Selector application for bulking control assumes that all filamentous organisms require raCOD for growth. This is certainly the case for some filamentous organisms, such as *S. natans*, Type 1701, *Thiothrix* spp., and Type 021N; however, there are filamentous organisms, such as *M. parvicella* and Type 0092, that apparently do not require raCOD for growth (Gabb and Jenkins, 1991). These filamentous organisms will not be controlled by selectors, which target raCOD removal prior to the main aeration zone.

Aerobic selectors utilize "kinetic" selection to promote substrate uptake and storage by floc-formers over filamentous organisms in the presence of oxygen. Since filamentous organisms proliferate under low substrate conditions, aerobic selectors typically apply a high process loading rate to the selector zone to create a high substrate concentration, which favors floc-formers. Aerobic selectors are most often staged to prevent breakthrough of raCOD to the main aeration zone during variations in influent characteristics and loading.

Anoxic selectors rely primarily on "metabolic" selection for raCOD removal, which is achieved by eliminating oxygen and allowing nitrate and nitrite to serve as the electron acceptors. Most filamentous organisms cannot use raCOD efficiently with nitrate/nitrite as the only available electron acceptor. This absence of a metabolic pathway means that filamentous organisms should be not be able to compete for raCOD in anoxic selector systems.

Similar to anoxic selectors, anaerobic selectors utilize metabolic selection by providing a zone where raCOD is taken up by phosphorus-accumulating organisms (PAOs) with an associated release of orthophosphate. Filamentous organisms that require raCOD are not able to compete with PAOs for raCOD under anaerobic conditions and will not survive in activated sludge systems unless raCOD leaks through to the main aeration zone. Enhanced biological phosphorus removal (EBPR) activated sludge systems rely on this same mechanism, which means that filamentous organisms requiring raCOD for survival will likely be eliminated from this type of system.

It is widely accepted that aerobic selectors should be staged to help ensure adequate raCOD removal during peak loading conditions, while providing the proper kinetic conditions (high substrate concentration) during periods of low raCOD loading. However, it is unclear whether anoxic and anaerobic selectors, which rely primarily on metabolic selection and not kinetic selection, should be staged for successful bulking control. A staged selector may improve raCOD removal efficiency by reducing short-circuiting issues that may be present in a single-stage selector.

A survey of 33 full-scale, long mean cell residence time (MCRT) EBPR activated sludge plants in South Africa (Blackbeard et al., 1987) showed that the most common filamentous organisms causing bulking were (in order of frequency of occurrence): Type 0092, Type 0675, Type 0041, *M. parvicella*, Type 0914, and Type 1851. *Thiothrix* spp., *S. natans*, Type 1701, Type 021N, and Type 0961 (filamentous organisms requiring raCOD) were rarely, if ever, identified in these EBPR activated sludges. This suggests that filamentous organisms common to long-MCRT EBPR activated sludges may not require raCOD and may not be controlled by selectors.

In contrast, a combined survey of 270 U.S. plants (Richard et al., 1982, Strom and Jenkins, 1984), identified the following bulking filaments (in order of frequency of occurrence): Type 1701, Type 021N, Type 0041, *Thiothrix* spp., *S. natans*, *M. parvicella*, and Type 0092. The difference in dominant bulking filament types relative to the South African plants is likely because U.S. plants did not incorporate EBPR designs and operate at long MCRTs when the surveys were conducted.

In this report, readily assimilable COD (raCOD) or slowly assimilable COD (saCOD) is used to describe substrates that are more commonly referred to as readily biodegradable COD (rbCOD) or slowly biodegradable COD (sbCOD), because assimilable is more accurate in the context of selector mechanisms. When a substrate is assimilated in activated sludge, it is taken up by a microbial cell and either stored within the cell or biodegraded for cell energy and growth. The biodegradable COD term, then, ignores COD uptake and storage. The continued use of "biodegradable COD" instead of the more accurate "assimilable COD" has been supported by the belief that any readily biodegradable substrate is also readily assimilable and vice versa. Whether a substrate is readily assimilable or readily biodegradable, however, is specific to the type of activated sludge cultured.

For example, acetate is considered both readily biodegradable and readily assimilable. In a perfect anaerobic selector system, however, all acetate is readily taken up by either PAOs or glycogen-accumulating organisms (GAOs) and stored in large intracellular molecules (granules). These storage molecules are later slowly biodegraded by the organism for energy and growth. Thus, in anaerobic selector systems, acetate is readily assimilable but slowly biodegradable.

The laboratory investigation in this study provides another example of acetate being slowly biodegradable in a particular activated sludge (Chapter 2.0). The synthetic sewage fed to the R2 bench-scale reactor initially contained a variety of substrates including acetate and other raCOD materials. High acetate and oxygen uptake rates were measured when acetate was fed to the R2 activated sludge in batch tests, indicating that acetate is both readily biodegradable (from the high oxygen uptake rate) and readily assimilable (from the high acetate uptake rate). Then, according to plan, the acetate and all other raCOD substrates were removed from the R2 feed. Following this feed change, the same batch test was repeated, but this time both acetate and oxygen uptake rates were dramatically lower, indicating that acetate was now a slowly biodegradable and slowly assimilable substrate (see Figure 3-2).

The sludge volume index (SVI), expressed in mL/g and based on the 30-min settled sludge volume (SSV_{30}) result, is commonly measured at treatment facilities to estimate sludge settleability. During bulking episodes, SVIs may rise above 150 mL/g—the typical upper limit for a well-settling sludge. Although the SVI test is widely applied, Dick and Vesilind (1969) reported that accuracy depends on mixed liquor solids concentration. Specifically, test accuracy fails when the SSV_{30} exceeds 250 to 300 mL/L. Stöbbe (1964) proposed testing sample dilutions such that the SSV_{30} yields a result between 150 and 250 mL/L, which he termed the diluted SVI (DSVI) method. Lee et al. (1983) reported that the DSVI test yielded the best correlation with total extended filament length compared to other techniques available for estimating sludge settleability. Merkel (1971) developed a relationship to convert measured SVI values to equivalent DSVI values (discussed further in Section 4.3.2) to allow more meaningful comparisons between settleability data collected from different facilities.

Although selectors may not control some filamentous organisms, their application in activated sludge processes has been widely accepted and often successful. Nonetheless, significant variation exists in process design and operating criteria and the degree of sludge settleability control. In fact, specific, consistent design guidelines and detailed design criteria are generally not available in the literature. Selector installations are commonly designed on an empirical basis, drawing heavily on design concepts and demonstrated performance at facilities with similar wastewater characteristics and process configurations. This approach has yielded inconsistent filamentous bulking control under similar design provisions, underscoring the challenging nature of selector design.

1.2 Project Objectives

The main objectives of this study were to:

- Investigate the mechanisms limiting the ability of a selector to control the growth of specific filamentous organisms known to cause bulking;
- Establish a project database of selector design and performance from a large pool of full-scale facilities (20-25) across the U.S.;

- Identify selector design and performance relationships for each of the three main selector categories (aerobic, anoxic, and anaerobic) based on the project database information collected; and
- Demonstrate implementation of a full-scale anaerobic selector at two wastewater treatment facilities and identify associated selector design and performance issues.

1.3 Project Scope and Approach

This study was organized into five main project tasks:

1. Literature Review
2. Laboratory Investigation
3. Initial Plant Screening Survey
4. Detailed Plant Investigations
5. Full-Scale Demonstration Projects

A literature search and review was conducted by Dr. H. David Stensel (University of Washington, Seattle). Specific topics of interest included filament type and occurrence in activated sludge processes, kinetic and metabolic substrate (raCOD) removal mechanisms, design and performance of full-scale selector installations, and current research efforts related to the control of specific filament types. A summary of the literature review report prepared by Dr. Stensel is presented in Chapter 2.0.

A bench-scale experiment was performed by Mr. Gang Xin at the University of Washington, Seattle, under the direction of Dr. Stensel, to investigate the ability of an aerobic selector to control two specific filament types—those that prefer raCOD (Type 021N, *Thiothrix*) and those that thrive on saCOD (*M. parvicella*, Type 0092). Since selector designs typically focus on raCOD removal, the ability of some filaments to use saCOD as their primary substrate may help to explain why some selector installations will consistently fail if the conditions promoting saCOD filament growth are present. The laboratory report prepared by Dr. Stensel has been summarized and is presented in Chapter 3.0.

An initial screening survey of 125 wastewater treatment plants was conducted to develop a comprehensive database of full-scale facilities encompassing a wide range of selector designs and performance. The screening level survey form (refer to Appendix A) was designed to be completed in a relatively short time period (approximately 15 minutes). The main goal of the initial survey was to collect the following basic plant information:

- Plant flow rate
- MCRT
- Nutrient removal requirements
- Aeration basin configuration
- Type of selector
- Bulking frequency
- SVI control achieved following selector installation

An additional function of the initial screening survey was to establish plant contacts and to identify candidate facilities willing to participate further. Following completion of the initial screening survey, 44 of the 125 facilities participated in a detailed plant investigation. As an initial step, each of the plants was asked to complete a detailed plant survey specifically designed for an aerobic, anoxic, or anaerobic selector installation (refer to Appendix A). The data requested included:

- Aeration system configuration and sizing
- Selector design, configuration, and sizing
- One year of plant operating and selector performance data
- Process schematic
- Technical reports related to selector design or evaluation

A complete list of the information requested from each of the facilities participating in the detailed plant investigation is provided in Chapter 4.0. The data collection process required numerous follow-up requests and discussions with plant contacts to verify the information provided and plant-specific issues. Based on the secondary process operating data provided, the following important design parameters were calculated in order to compare selector design and performance between facilities, including:

- Selector and aerated MCRT
- Contact (or floc) loading
- Selector initial contact zone (ICZ) food-to-microorganism (F/M) loading
- Nominal total selector and ICZ hydraulic residence time (HRT)
- Total selector and ICZ HRT (with recycle)
- Effective number of selector stages
- Average and 90th percentile SVI
- Average and 90th percentile Merkel-corrected DSVI (Merkel, 1971)
- Percent of SVI values greater than 150 mL/g

The main goal of the detailed plant investigation was to evaluate typical selector design and operating parameters against settling performance at a large number of facilities equipped with selectors. A summary of the initial and detailed plant investigation results is presented in Chapter 4.0.

This study included demonstration of full-scale anaerobic selectors at two wastewater treatment facilities—the East Bay Municipal Utility District (EBMUD) Main Wastewater Treatment Plant (MWWTP) in Oakland, Calif., and Orange County Sanitation District (OCSD) Plant No. 1 in Fountain Valley, Calif. The goal of this work was to provide municipalities with key information necessary for successful selector implementation at their facilities by highlighting process considerations and issues. Chapter 5.0 summarizes the main findings and conclusions of the two full-scale anaerobic selector demonstration projects conducted by EBMUD and OCSD.

This study also provides a computerized selector diagnostic tool to assist troubleshooting and designing selector systems. This computerized selector diagnostic tool has been copied to a CD-ROM, which has been placed in the inside back cover of this report. Documentation for this software is provided in Appendix F.

CHAPTER 2.0

LITERATURE REVIEW SUMMARY

2.1 Background

A review of available literature on selector processes was conducted early in the project to assist with the planned bench- and full-scale selector evaluation tasks. The main goals of the literature review were as follows:

- Identify the operating and metabolic conditions that favor the growth and occurrence of specific filamentous bulking organisms in activated sludge systems, and
- Summarize the design and performance of pilot- and full-scale selector applications.

A summary of the literature review, which covers the period up to January 2004, is presented in this section.

2.2 Selector Application

Selectors are process tank configurations installed prior to the main aeration basin in activated sludge systems to favor the growth of floc-forming bacteria and minimize the growth of filamentous organisms. The principle of selector operation is to direct the substrate (carbon source) to the preferred floc-forming microorganisms, thereby minimizing its availability for the growth of filamentous organisms. The application of selectors to control filamentous bulking has generally resulted in the addition of relatively short hydraulic detention time initial contact tanks, in which RAS and influent wastewater are mixed prior to the main aeration zone. Selectors are typically classified as aerobic, anoxic, or anaerobic based on the environmental conditions present within these tanks. In some cases, it is possible to have aerobic/anoxic or anoxic/anaerobic conditions within the floc or at different locations in the same selector tank.

The effectiveness of selector processes in controlling SVIs has been variable. Parker et al. (2003) found that six of 21 facilities (29%) equipped with anoxic or anaerobic selectors experienced 90th percentile SVI values above 150 mL/g, which is a typical upper limit for well-settling sludges. Possible factors contributing to poor selector performance include selector design, main aeration tank design, plant operating conditions, and wastewater characteristics.

2.3 Filament Type and Occurrence

Significant efforts have been made by researchers to identify the many different types of filamentous microorganisms and the conditions favoring their growth and proliferation in activated sludge systems. Eikelboom (1975) first developed and applied a systematic approach for characterizing filament types in various activated sludge plants in Europe. Characterization techniques and the database of information were advanced further in work by Jenkins et al. (1984) and Wanner and Grau (1988). These methods relied primarily on microscopic examination of filament morphology, size, and staining characteristics. Recent advances in molecular biology have enabled some researchers to base filament identification techniques on DNA or rRNA composition.

The factors influencing the type and occurrence of filamentous organisms can be summarized into three categories: 1) wastewater characteristics, 2) process configuration designs, and 3) operational conditions (Jenkins et al., 2004). A summary of conditions favoring the growth of specific filament types is presented in Table 2-1, which was compiled from Jenkins et al. (2004) and Eikelboom (2000).

Table 2-1. Summary of Occurrence Conditions of Commonly Observed Filamentous Organisms.

Filament Type	Substrate			Sulfide	Low DO	F/M Loading[4]		
	raCOD[1]	LCFA[2]	saCOD[3]			Low	Med	High
Type 0041			✓			✓		
Type 0675			✓			✓		
Type 1851	✓					✓		
Type 0092			✓			✓		
Type 0803			✓			✓		
M. parvicella		✓	✓		✓	✓		
Type 0961	✓					✓	✓	
Nocardioforms	✓		✓			✓	✓	
Type 0914	✓		✓	✓		✓	✓	
Type 021N	✓			✓		✓	✓	
N. limicola	✓		✓			✓	✓	✓
H. hydrossis	✓				✓	✓	✓	✓
Type 1701	✓				✓		✓	✓
Type 0411	✓						✓	
Thiothrix sp.	✓			✓			✓	✓
Type 1863	✓							✓
S. natans	✓				✓		✓	✓

Notes:

[1] raCOD = readily assimilable COD

[2] LCFA = long-chain fatty acids; filament types in the LCFA category are also in the saCOD category

[3] saCOD = slowly assimilable COD

[4] Low F/M: <0.2; medium F/M: 0.2–0.7; high F/M: >0.7 kg BOD_5/kg MLSS-d

2.3.1 Wastewater Characteristics

Domestic and industrial wastewaters contain organic substrates with wide ranges of composition and form, including carbohydrates, proteins and other organic nitrogen compounds, short-chain fatty acids (SCFAs), and LCFAs. In domestic wastewaters, a portion of the influent COD is soluble and readily assimilable and is often referred to as raCOD[1]. The raCOD fraction may range from 15%–40% of the total influent COD (Tchobanoglous et al., 2003) in municipal wastewaters, depending on plant location, collection system length and slope, industrial loading, and temperature. The remaining assimilable COD consists of particulates (colloidal and suspended solids) and dissolved fats, oil, and grease.

The presence of raCOD substrates, especially SCFAs, is critical for the growth of many filamentous bulking organisms, as shown in Table 2-1. The role of raCOD in the growth of certain types of filaments, including Type 021N (van Niekerk et al., 1987, Richard et al., 1985, Kampfer et al., 1995, Kohno, 1989, Pernelle et al., 2001, Andreasen and Nielsen, 1997), *Thiothrix* (Tandoi et al., 1994, Pernelle et al., 2001, Nielsen et al., 2000), *H. hydrossis* (Pernelle et al., 2001), and *S. natans* (Contreras et al., 2000) has been demonstrated in pure culture and in

[1] Refer to raCOD discussion on Page 1-3 in Chapter 1.0.

activated sludge experiments. Although the influent wastewater is the most common source of raCOD, it may also be provided by the hydrolysis of influent particulate COD. Ekama and Marais (1986) hypothesized that hydrolysis products are not completely consumed within the flocs and may eventually be released into bulk solution, where they are available for both floc-formers and filamentous organisms. Particulate hydrolysis can also occur in equalization basins or primary sedimentation tanks and has been related to elevated filament growth in systems operating without selectors (Jenkins et al., 2004).

Successful selector designs typically target the removal of raCOD, since this is the primary substrate source for some of the most common bulking filaments (refer to Table 2-1). However, it is important to note that certain filament types do not rely on raCOD, which may explain why in some cases selectors are not effective in preventing sludge bulking. For example, some filament types, as well as floc-forming organisms, rely on LCFAs (oleic acid) for growth. Several researchers (Gabb et al., 1991, Nielsen et al., 2002, Andreasen and Nielsen, 2000) indicated that *M. parvicella* has a considerable capacity for LCFA uptake under aerobic, anoxic, and anaerobic conditions. Other saCOD substrates (starch) may also be used by filamentous organisms in CMAS systems at low-to-moderate F/Ms.

Given that selectors are typically designed for raCOD removal, some researchers have raised concerns over the fate of saCOD in the main aeration basin and conditions that may favor substrate uptake by filamentous organisms. Kappeler and Gujer (1993), however, claimed that the particulate saCOD hydrolysis products in the activated sludge process favor floc-forming organisms rather than filamentous organisms because the particulates are captured within the floc and there is a short distance between the site of raCOD production from particulates and the site of consumption.

Some filamentous organisms, such as *Thiothrix*, *Beggiatoa*, and Type 021N, use sulfide present in the influent wastewater to gain energy and low molecular weight organics as the carbon source. Sulfide may be present at sufficient concentrations to promote filament growth if the influent wastewater is septic or if significant sulfate reduction occurs somewhere in the treatment process (anaerobic zones, primary clarifiers, sludge thickeners).

Nutrient deficiencies, such as limited nitrogen (N) or phosphorus (P), can favor the growth of several filamentous organisms. Type 021N and *Thiothrix* grow well under N deficiency, while *S. natans* and *H. hydrossis* can grow under P deficiency conditions (Jenkins et al., 2004).

2.3.2 Activated Sludge Process Designs

Early researchers began to understand the relationship between sludge settleability and aeration basin configuration by observing SVIs at full-scale facilities. Chambers and Tomlinson (1982) demonstrated that CMAS systems often had higher SVIs than those systems operated with a substrate gradient (conventional plug flow reactors with long narrow tanks, staged reactors, fill-and-draw processes). Similarly, Pasveer (1969) observed the significance of feeding and mixing characteristics in oxidation ditch systems by noting an increase in settleability problems as early fill and draw designs were converted to completely mixed, continuous-flow systems. In sequencing batch reactors (SBRs), Martins et al. (2003a) showed that fill time had an important effect on sludge bulking. Shorter fill times of the same feed volume produced well-settling sludge, while longer fill times produced a bulking sludge. Jenkins et al. (1984) indicated that an activated sludge system operated with a feast-famine cycle could produce a selector effect to favor floc-formers. In a feast-famine cycle, a short filling period in an SBR or a small initial contact zone in a continuous-feed activated sludge process can create a food-rich phase called the "feast" phase, which is followed by a "famine" phase characterized by a sufficiently long aeration time to fully oxidize the food gained in the "feast" phase. These observations on reactor configuration and the role of mixing regimes eventually led to the development of selector technologies.

2.3.3 Operational Conditions

The growth of specific bulking filaments may be controlled by activated sludge process operating conditions, including DO levels, temperature, and pH. Excessive growth of many filamentous organisms (*S. natans*, Type 1701, *H. hydrossis*, and *M. parvicella*) is associated with low DO conditions (DO <2 mg/L in bulk solution) (refer to Table 2-1). In activated sludge systems, low DO is a relative term because the DO levels in the floc are affected by the bulk liquid substrate and DO concentrations. At higher F/Ms, the raCOD concentration is higher, resulting in a higher oxygen uptake rate. A higher bulk liquid DO concentration is thus needed to maintain an aerobic floc as the reactor F/M loading is increased. Palm et al. (1980) recommended a DO operating line as a function of F/M (summarized in Table 2-2). An F/M higher than the number provided in Table 2-2 under the corresponding DO concentration is sufficient to promote sludge bulking. However, the work only addressed the system response at higher DO concentrations (>1.0 mg/L) and did not address effectiveness at lower DO concentrations (<0.5 mg/L) and higher process loadings typically present in a selector. Moreover, these recommendations only apply to CMAS systems.

Table 2-2. Combinations of F/M and Aeration Basin DO Level Above Which Low DO Bulking Does Not Occur in Completely Mixed, Continuously Fed Aeration Basins (adapted from Palm et al., 1980).

F/M [kg COD removed/ (kg MLVSS·d)]	Aeration Basin DO Level (mg/L)
0.2	1.0
0.4	2.0
0.6	2.6
0.8	3.6
1.0	4.2
1.2	4.9
1.4	5.7

Martins et al. (2003b) agreed that the DO content should be related to the F/M in activated sludge systems, but they proposed much lower DO concentrations than Palm et al. (1980). High F/M [>20 kg COD/(kg MLVSS·d)] in the initial zone of an activated sludge process requires a DO concentration as high as 2.0 mg/L. The DO concentration should be maintained above 1.0 mg/L when the F/M is 10–13 kg COD/(kg MLVSS·d). In addition, the presence of microaerophilic conditions (DO range of 0.1–0.5 mg/L) in the anoxic stage of laboratory anoxic-aerobic systems was reported by Martins et al. (2004a) to have caused poor sludge settling. Although elevated DO concentrations may be used to favor the growth of floc-formers, it does not guarantee a lack of filamentous growth since other factors, such as reactor configuration, are also important. Also, as Donkin (1999) suggests for Type 0411, some filaments prefer a high DO concentration environment.

Within the normal temperature range in activated sludge basins (8-25°C), the growth rate of filamentous organisms increases more rapidly with temperature increases than the growth rate of floc forming organisms (Jenkins et al., 2004, Krishna and van Loosdrecht, 1999). Morgan-Sagastume and Allen (2003) also pointed out that increasing the temperature from 35-45°C deteriorated the sludge settleability by promoting filament growth. An exception is *M. parvicella*, which can reduce the sludge settleability more at low temperatures (<15°C) than at high temperatures. This is thought to be caused by a smaller temperature effect on its specific growth rate, compared to other organisms, as temperature is decreased (Knoop and Kunst, 1998, Soddell and Seviour, 1995). Eikelboom (2000) noted that the effects of temperature on filaments that use LCFA in oxidation ditches operated at a high MCRT were as follows: *M. parvicella* is dominant at temperatures below 15°C, and Type 0092 is dominant at temperatures above 15°C. A similar switch was observed for *Thiothrix I* and *II*, with *Thiothrix I* being observed at temperatures above 20°C and *Thiothrix II* being observed at temperatures below 20°C (Donkin, 1999).

A summary of operating conditions commonly identified with specific filamentous organisms is presented for reference in Appendix B.

2.4 Most Common Filaments at Wastewater Treatment Plants

Using the microscopic examination techniques described by Jenkins et al. (2004) and Eikelboom (2000), specific filamentous organisms have been identified in mixed liquor samples from full-scale activated sludge plants in various countries. Table 2-3 is a summary of the most common filaments observed at wastewater treatment plants.

Table 2-3. Most Common Filamentous Organisms Reported at Wastewater Treatment Facilities. (ranked in descending order).

Country	Number of Plants	Plant Types	Most Common Filaments	Reference
United States	270	Various	[1] Nocardioforms–31%, Type 1701–29%, Type 021N–19%, *Thiothrix*–16%	Jenkins et al. 2004
The Netherlands	93	Mostly oxidation ditches	[1] *M. parvicella*–58%, Type 0041	Eikelboom, 2000
Denmark	81	BNR plants EBPR and MLE	[1] *M. parvicella*–61%, Type 0041–28%, *N. limicola*–10%	Eikelboom, 2000
Greece	11	Various	[1] *M. parvicella*–75%, Type 0803–12%	Eikelboom, 2000
Czech Republic	86	Various	*M. parvicella*, *N. limicola*, Type 0092, Type 0803	Wanner et al., 1998
South Africa	33	BNR plants	[2] Type 0092–82%, Type 0675–45%, Type 0041–39%, *M. parvicella*–33%, Type 0914–33%	Blackbeard et al., 1988
South Africa KwaZulu-Natal	6	Various	Nocardioforms, Type 0041, Type 0675, Type 1851, Type 021N	Lacko et al., 1999
Italy	112	Various	[2] *M. parvicella*–61%, Type 0041–52%, *N. limicola*–40%, H. hydrossis–33%, Type 021N–32%	Madoni et al., 2000
Argentina	10	Municipal and industrial	Type 021N, *Thiothrix I*, *S. natans*, *M. parvicella*	Di Marzio, 2002
Thailand	6	Domestic and industrial	Type 021N, Type 1701, Type 0092, Type 0041	Mino, 1995

Notes:
[1] Reported as percent of all occurrences for all plants surveyed.
[2] Reported as occurrence in percent of plants surveyed.

2.5 Readily Assimilable Substrate Removal Mechanisms

Since raCOD can be taken up much faster by activated sludge than saCOD, the strategies favoring its uptake by floc-formers need to focus on the initial contact zone (ICZ) of the activated sludge system, while strategies for favoring saCOD uptake by floc-formers must consider the entire aeration time. In this section, selection of floc-formers based on kinetic, storage, and diffusion-based processes for removing raCOD is discussed.

2.5.1 Substrate Kinetic Selection Based on Growth Kinetics

Chudoba et al. (1973b) proposed that the selection mechanism in a staged selector configuration was related to substrate utilization kinetics; specifically, floc-formers had a higher specific growth rate at higher raCOD concentrations. Using Monod equation (Monod, 1949) growth kinetics, Chudoba hypothesized that filamentous organisms thrive in single-stage, well-mixed reactors with low substrate levels. This work emphasized the need to create a substrate concentration gradient by staging the selector zone in aerobic selectors for effective bulking control.

2.5.2 Substrate Storage Mechanisms and Kinetic Selection

When bacteria are able to assimilate raCOD into storage without depending solely on oxidation and synthesis, more substrate consumption at a given DO level can occur. This mechanism may provide the microorganism with a faster apparent growth rate at high substrate levels. This section describes storage selection processes under aerobic and anoxic conditions.

2.5.2.1 Substrate Storage Under Aerobic Conditions

Given a sudden increase in raCOD concentration, microorganisms adapt by increasing their substrate uptake rate and/or substrate storage capacity (Daigger and Grady, 1982a and 1982b). Since the uptake and storage response occurs more rapidly than the growth rate response, organisms that are able to store raCOD during the feast period and subsequently consume the stored substrate for growth during the famine period will have a greater competitive advantage. Floc-forming organisms were reported to have a higher storage capacity (Grau et al., 1982) and an associated competitive advantage in processes with feast-famine cycles. Axenic (van den Eynde et al., 1983) and mixed (Chudoba et al., 1982 and 1985) culture studies have shown a higher storage response for floc-forming organisms.

2.5.2.2 Substrate Storage Under Anoxic Conditions

In an anoxic environment, nitrate-reducing bacteria can form internal storage polymers for use as a substrate source during the famine period. Some filamentous organisms are able to use nitrate as an electron acceptor, such as *M. parvicella* (Tandoi et al., 1998), *S. natans* (Pellegrin et al., 1999), *Thiothrix* (Williams and Unz, 1985), Type 021N (Williams and Unz, 1985), and Type 1851 (Kohno et al., 2002); however, denitrification rates for filaments (*Thiothrix*, Type 021N) are reportedly much lower than for floc-formers (Shao and Jenkins, 1989 and Dionisi et al., 2002). Furthermore, storage capabilities have only been shown for two filament types to date—*N. limicola II* (Dionisi et al., 2002) and *M. parvicella* (Andreasen and Nielsen, 1997)—compared to the widespread substrate storage ability among floc-forming organisms; therefore, the denitrification-nitrification process should effectively control filaments. Selection under anoxic conditions does not necessarily require alternating anoxic/aerobic conditions; raCOD may be taken up and metabolized directly under anoxic conditions.

2.5.3 Metabolic Selection

In EBPR systems, PAOs and glycogen-accumulating organisms (GAOs) exhibit a different metabolic pathway than non-PAO/GAOs. Under anaerobic-aerobic cycling, raCOD may be taken up and stored as poly-β-hydroxybutyrate (PHB) in the anaerobic zone by PAOs and GAOs. PAOs use the hydrolysis of stored high energy inorganic polyphosphate and the fermentation of stored glycogen to provide energy for uptake and storage. GAOs take up and store PHB using energy generated by the fermentation of stored glycogen. During the aerobic period, the stored carbon products are oxidized by PAOs, orthophosphate is taken up to resynthesize the inorganic polyphosphate, and stored energy is provided by oxidation. To date, no filamentous organisms have been identified as having substrate uptake mechanisms similar to PAOs or GAOs in anaerobic systems.

2.5.4 Diffusion-Based Selection

Martins et al. (2003a and 2003b) proposed a diffusion-based selection theory, which states that filamentous organisms have a higher specific growth rate at low substrate levels because they have easier access to bulk liquid substrate. Since filamentous organisms grow in one or two directions and floc-formers grow in three directions, filaments would appear to have better access to raCOD, and their growth rate is not limited by a lower diffusion rate into the floc as for floc-formers.

2.6 Slowly Assimilable Substrate

Gabb et al. (1991) noted that an aerobic selector did not control the growth of what was referred to as "low F/M filaments." In addition to being slow-growing bacteria, these organisms also preferred saCOD rather than the raCOD that floc-forming bacteria out-competed filamentous bacteria for in selector applications. This condition allows filamentous organisms that prefer saCOD to proliferate in systems that are successful in removing raCOD in the selector zone.

Lipids are present in all domestic wastewater streams and contribute a significant portion (25%-30%) of the organic substrate (Quemeneur and Marty, 1994 and Raunkjaer et al., 1994). The major lipid fraction is present as triglycerides (TAG) and a minor amount is free LCFA (Quemeneur and Marty, 1994). Under batch conditions, Dueholm et al. (2001) found TAG cannot be consumed directly by the bacteria in activated sludge but must first be hydrolyzed to LCFA. They also noted that the observed yield coefficient for oleic acid degradation was in the same range as the yield coefficient for acetate utilization in activated sludge systems. They further showed that LCFA could easily be consumed under aerobic and anoxic conditions. The rapid removal of LCFA in activated sludge was believed to be by biosorption (Hwu et al., 1998 and Tsezos and Bell., 1989) or storage (Nielsen et al., 2002). Under anaerobic-aerobic conditions, Nielsen et al. (2002) observed that a large fraction of the removed lipids were neutral lipids, which may be present as storage products, rather than polar lipids, which include membrane phospholipids. In addition, they also found that the polar lipid fraction became larger in the aerobic stage of the anaerobic-aerobic systems. Hence, they proposed that some organisms in activated sludge could take up and store LCFA under anaerobic conditions and subsequently use the storage material for growth with oxygen or nitrate as the electron acceptor.

M. parvicella is relatively hydrophobic, suggesting that hydrophobic substrates such as LCFA will preferentially adsorb to *M. parvicella* compared to the other organisms present in the same study (Nielsen et al., 2002). Uptake and storage of LCFA under anaerobic conditions provides an effective competitive advantage for *M. parvicella* against those organisms (both floc-former and filaments) that can only take up LCFA under aerobic conditions (Andreasen and Nielsen, 2000 and Nielsen et al., 2002). When a decrease in water temperature reduces the solubility of LCFA, *M. parvicella* may maintain a similar LCFA uptake ability, which provides a further benefit against other filaments utilizing LCFA. This has been proposed as an explanation as to why *M. parvicella* is more common in activated sludge systems at lower temperatures (Andreasen and Nielsen, 2000).

2.7 Selector Processes and Designs

The primary focus of selector processes and designs is to provide an environment that favors the growth of floc-formers over filamentous organisms. A major concern is the removal of raCOD substrates in the selector zone; however, some filamentous organisms can survive on saCOD substrates and may not be controlled by selector processes. Selectors are categorized as aerobic (oxygen as electron acceptor), anoxic (nitrate/nitrite as electron acceptors), or anaerobic (no oxygen or nitrate/nitrite present). This section summarizes available design guidelines and recommendations for each of these selector classifications.

2.7.1 Substrate Removal

In order to control filamentous bulking, Shao and Jenkins (1989) recommended that the sCOD concentration leaving the selector be <60 mg/L and demonstrated that raCOD was effectively zero under this condition. Although this value is based on data collected for an anoxic selector, it may be broadly applied to all selector types. Chudoba and Wanner (1987) recommended that the selector zone should remove 80% of the removable COD, which is defined as the difference between secondary influent and effluent sCOD.

2.7.2 Selector Staging and Configuration

It is widely accepted that aerobic selectors should be staged to provide the proper kinetic conditions favoring substrate uptake and storage by floc-formers over filaments; however, the need to stage anoxic and anaerobic selectors, which rely on metabolic rather than kinetic selection, is subject to debate. Selector staging in anoxic and anaerobic selector systems may improve raCOD removal by providing both kinetic and metabolic pathways for floc-formers, yielding more efficient uptake of available substrate than for a single-staged reactor. It is conceivable that the higher substrate uptake efficiency in the initial selector stages would result in a lower raCOD concentration leaving the selector zone than for a single-stage selector of the same volume. Staged selector designs may also provide greater reliability under peak or variable loading conditions. Further, the higher substrate concentrations present in the initial stages may encourage more rapid substrate uptake and storage by floc-formers.

Jenkins et al. (2004) recommended that all three selector types be designed with at least three stages, sized at 25%, 25%, and 50% of the total selector volume, respectively. Similarly, Albertson (2005) recommended that all selectors be staged to create an F/M gradient. Staged selectors may also increase selector loading capacity, and during peak raCOD loadings, less raCOD should leak to the main aeration zone relative to a single-stage selector.

Wanner (1994) noted a number of studies in which good SVI control was achieved in staged anaerobic selector systems. These include studies by Watanabe (in Wanner, 1994) using a three-stage anaerobic zone in a laboratory study and by Daigger and Nicholson (in Wanner, 1994) using a six-stage anaerobic zone followed by a four-stage oxic zone in a pilot plant study in Fayetteville, Ark. Albertson (in Wanner, 1994) also reported SVI improvements with a three-stage anaerobic/anoxic zone, followed by an aeration zone in a full-scale wastewater treatment plant in Newark, Ohio. Wanner (1994) also reported, however, that a single-stage anaerobic selector successfully controlled filamentous bulking when properly sized.

2.7.3 Selector Design Loadings

Table 2-4 presents a comparison of recommended loadings for different selector types by Albertson (2005), Jenkins et al. (2004), and Chudoba and Wanner (1988). The recommendations for aerobic selectors are based on providing proper conditions for kinetic selection while avoiding excessive loading conditions that may promote viscous bulking. Albertson (2005) noted that for the Phoenix, Arizona 23rd Avenue WWTP bulking occurred when the ICZ F/M loading was >9.0 kg BOD_5/(kg MLSS·d) due to a large loading increase from a cheese production facility. Chudoba's (1973b) earlier work on staged aerated selectors found that SVI values of <100 mL/g were achieved when the ICZ F/M was 2.3–4.4 kg BOD_5/(kg MLSS·d). Albertson (2005) recognized the importance of describing ICZ loading on either a soluble BOD_5 ($sBOD_5$) or sCOD basis. The design COD loading values recommended by Albertson (2005) in Table 2-4 are related to sCOD loadings by multiplying by 0.5.

Table 2-4. Recommended Design F/M Loadings for Staged Selectors.

Stage No.	F/ΣM [kg COD/(kg MLSS·d)][1] Aerobic Selector High DO	Aerobic Selector Low DO	Anoxic Selector	Anaerobic Selector	Reference
1	16	12	10-12	12	Albertson (2005)[2]
2	8	6	5-6	6	
3	4	3	1.5-3	3	
1	--	12	6	6	Jenkins et al. (2004)
2	--	6	3	3	
3	--	3	1.5	1.5	
1	--	12	--	--	Chudoba and Wanner (1988)
2	--	6	--	--	
3	--	4	--	--	
4	--	3	--	--	

Notes:

[1] ΣM is the mixed liquor mass in the selector stage plus that in prior stages.

[2] Actual recommended design guidelines were reported on an sCOD basis and converted to a COD basis as follows: COD = sCOD x 2.

For single-stage anoxic selectors, Marten and Daigger (1997) recommended a selector F/M of ≤1.0 kg BOD_5/(kg MLSS·d) for temperatures ≤18°C and ≤1.5 kg BOD_5/(kg MLSS·d) for temperatures >18°C, while the anoxic MCRT should be 1–2 d. Grady et al. (1999) recommended an anoxic MCRT of 1.0 d at temperatures >20°C and 1.5 d at temperatures <17°C. Jenkins et al. (2004) recommended a total selector HRT for anaerobic selectors of 0.75–2.0 h.

The floc loading or contact loading may be an important factor for selector applications with high strength wastewaters. Floc loading is the ratio of sCOD mass to MLSS mass applied to the ICZ. The concept of a contact loading limitation recognizes that the mixed liquor has a finite capacity for uptake of the influent-soluble substrate. Thus, at very high contact loadings, a portion of the influent sCOD cannot be consumed, and will be available for metabolism by filamentous bacteria in the subsequent main aeration zone. The recommended limiting value for the contact loading by Albertson (2005) is 0.1 g sCOD/g MLSS. This value basically assumes that the mixed liquor has the capacity for uptake and storage of substrate at about 10% of its dry weight. Albertson points out that for sCOD concentrations typical of municipal wastewaters

(125–250 mg/L). the initial floc loading may be in the range of 0.025–0.06 g/g after combining the influent flow and the return sludge flows, assuming MLSS concentrations of 3,000–3,500 mg/L. At this floc loading condition, the ICZ loading may be in the range of 4.0–6.0 kg BOD_5/(kg MLVSS·d).

2.8 Full-Scale Selector Operation and Performance

This section presents selector operating and performance conditions for 36 full-scale facilities operating with aerobic, anoxic, or anaerobic selectors.

2.8.1 Aerobic Selectors

Table 2-5 includes operating and performance data for 11 plants equipped with aerobic selectors with seven facilities reporting improved bulking control following selector installation. Factors contributing to selector failure at the remaining facilities include nutrient deficiency (Mill A1, Mill C), growth of *M. parvicella* due to low selector DO levels (Northside, Okla.), and growth of *M. parvicella* due to LCFAs (Cologne-Langel, Germany).

Daigger and Nicholson (1990) evaluated two aerobic selector plants achieving different levels of SVI control—Upper Occoquan Sewage Authority (UOSA), Va. and Northside, Okla. The UOSA selector design had an ICZ F/M loading of 14.9 kg BOD_5/(kg MLSS·d), which is higher than values recommended by Albertson (2005), and an HRT of 14 min. The Northside selector design had an ICZ F/M loading of 3.2 kg BOD_5/(kg MLSS·d) and a very short selector HRT. Testing at both facilities indicated that approximately 60% of $sBOD_5$ was removed by the selectors. Following selector installation, UOSA and Northside achieved SVIs of 74 and 152 mL/g, respectively, with apparent control of *M. parvicella* at UOSA but not at Northside. Daigger and Nicholson suggested that Northside may not have performed as well because the mechanical surface aerators may have allowed low DO concentration zones in the aeration basins and associated *M. parvicella* growth. In contrast, the two-stage aeration system at UOSA included both diffused air followed by mechanical surface aeration. DO concentrations in the aerobic selectors at UOSA and Northside were reported as 2.0 and 1.0 mg/L, respectively.

Duine and Kunst (2002) found that a three-stage aerobic selector with a total HRT of 20 min produced an SVI of 100 mL/g, while a single-stage selector with an HRT ranging from 15–40 min yielded poor SVI control. Based on the presence of raCOD filaments (Type 021N and Type 0961) in the single-stage selector, Duine and Kunst (2002) suggested that the low substrate concentration present in the selector favored the growth kinetics for filamentous organisms. However, no selector DO concentration data was presented to further support this point. Similarly, Rensink and Donker (1991) showed that a six-stage selector controlled the growth of *S. natans* and produced an SVI of 100 mL/g, while a single-stage selector yielded SVIs in the range of 200–400 mL/g with poor control of *S. natans*.

Table 2-5. Summary of Full-Scale Aerobic Selector Operating and Performance Conditions.

Plant	Mancasale, Italy	Pulp and Paper Mill Plants, USA				
		Mill A1	Mill A2	Mill C	Mill E	Mill F
Reference	Madoni and Davoli, 1997	Marshall and Richard, 2000				
Type of Wastewater	--	Sulphite	Thermo-mechanical pulp (TMP)	De-inking	TMP, unbleached sulphite	Groundwood, sulphite
Flow, m^3/d	9,600	--	--	--	--	--
Number of Selector Stages	1	3	3	1	2	1
Selector Volume as Percent of Total Reactor Volume, %	5	--	--	--	--	--
Aerobic HRT, h	4.75	--	--	--	--	--
Type of Aeration	--	Jet aerator	--	--	--	--
Temperature, °C	15	--	--	--	--	--
Aerobic MCRT, d	7.4-9.4	--	--	--	--	--
Aeration Basin F/M, kg BOD_5/(kgMLSS·d)	0.26	--	--	--	--	--
Selector ICZ F/M, kg BOD_5/(kgMLSS·d)	24	6.0-12.0	1.5-6.0	1.5-2.5	2.4-3.4	3.8-4.4
Selector COD Reduction, %	--	--	--	--	--	--
SVI without Selector, mL/g	160-620	--	--	--	--	--
SVI with Selector, mL/g	120-400	40-160	30-130	60-130	50-250	50-300
Bulking (B)/Foaming (F) without Selector	F	B	B	B	B	B
Bulking (B)/Foaming (F), with Selector	--	--	--	--	--	B
Dominant Filaments without Selector	*M. parvicella*	--	--	--	--	--
Dominant Filaments with Selector	*M. parvicella*	*N. limicola III*, *Thiothrix II*, Type 0914	*N. limicola III*, *Thiothrix II*, Type 0914	*H. hydrossis*, *Thiothrix I* *N. limicola II*	*Thiothrix I*	--
Effect of Selector Installation	Good	Unsteady	Initially poor, gradual improvement	Unsteady	Good, occasional bulking	Initially poor, then positive

Table 2-5. Summary of Full-Scale Aerobic Selector Operating and Performance Conditions (cont'd).

Plant	Cologne-Langel, Germany	Upper Occoquan Sewage Authority (UOSA), VA	Northside, Oklahoma City, OK	Leopoldsdorf Beet Sugar Plant, Austria	Potato-starch Plant, Austria
Reference	Lebek and Rosenwinkel, 2002	Daigger and Nicholson, 1990		Prendl and Kroiss, 1998	Nikolavcic and Svardal, 2000
Type of Wastewater	--	Domestic, industrial	Domestic, industrial	Beet sugar mill	Food-starch
Flow, m^3/d	18,900	102,000	105000	115,200	2,700-4,300
Number of Selector Stages	3	3	1	2-4 (equal volume)	3 (equal volume)
Selector Volume as Percent of Total Reactor Volume, %	2	2.6	4.9	2	4
Aerobic HRT, h	11	--	--	3.33	8.58
Type of Aeration	--	Diffused air followed by mechanical aeration	Surface mechanical aeration	--	--
Temperature, °C	--	--	--	18-28	9-27
Aerobic MCRT, d	--	--	--	5-8	6-13
Aeration Basin F/M, kg BOD_5/(kgMLSS·d)	--	--	--	0.2-0.3 (COD)	--
Selector ICZ F/M, kg BOD_5/(kgMLSS·d)	19.5 as COD	14.9	3.2	28 (COD)	2 (TOC)
Selector COD Reduction, %	--	45 (sCOD), 60 ($sBOD_5$)	60 ($sBOD_5$)	70	--
SVI without Selector, mL/g	--	up to 500	up to 500	--	300-500
SVI with Selector, mL/g	150-170	74	152	--	150-200
Bulking (B)/Foaming (F) without Selector	B/F	B/F	B/F	B	B
Bulking (B)/Foaming (F), with Selector	B/F	--	--	--	--
Dominant Filaments without Selector	*M. parvicella*	*M. parvicella*	*M. parvicella*	Type 0041, Type 021N, Type 1701	Type 021N
Dominant Filaments with Selector	*M. parvicella*	--	*M. parvicella*	--	Type 021N
Effect of Selector Installation	Poor	Good	Poor	Good	Good following adjustments

2.8.2 Anoxic Selectors

A summary of operating and performance data for 14 facilities equipped with anoxic selectors is provided in Table 2-6. Installation of an anoxic selector improved sludge settling performance at seven of the facilities; however, the other seven plants experienced no or little process improvements. The lack of improvement at these plants was due primarily to occasional high influent loading and/or cold temperatures, which allowed raCOD to break through to the main aeration basins (Beloit, Wisc.; Green Bay North and South, Wisc.; Italy B and D). These conditions, coupled with the presence of LCFAs and low DO concentrations, allowed filamentous organisms to proliferate in the activated sludge system with no significant improvement in SVIs following selector installation.

Installation of an anoxic selector at the Phoenix, Arizona 23rd Avenue Wastewater Treatment Plant reduced SVI levels from 275 to 100 mL/g. The selector was designed to provide three stages with HRTs of 15 min, 15 min, and 30 min, respectively, and an ICZ loading of 8 kg BOD_5/(kg MLSS·d). This design was completed in 1991 and is not consistent with current recommendations provided by Albertson (2005), as summarized in Table 2-4.

Davoli et al. (2002) reported on improved SVI control following installation of a small anoxic selector stage ahead of an existing anoxic/anaerobic system at four facilities in Italy (see Table 2-6). Prior to selector installation, each of the systems had a denitrification tank with an HRT ranging from 2.0–5.0 h. Low influent BOD_5 loading (43–78 mg/L) and relatively low DSVIs were reported, although Plant D had an average DSVI of 210 mL/g. The presence of *M. parvicella*, Type 0914, and Type 0041 filaments was reported. Small anoxic selector stages with HRTs ranging from 0.3–1.8 h were installed. Although the ICZ F/M loading was only 0.8 kg BOD_5/(kg MLSS·d) for Plant D, the DSVI was reduced from 210 to 108 mL/g; however, *M. parvicella* persisted.

Marten and Daigger (1997) reviewed the performance of four anoxic selector systems, including three single-stage and one three-stage configuration, with ICZ F/M loadings of 0.6–1.9 kg BOD_5/(kg MLSS·d). The anoxic selectors at the two Green Bay facilities and Landis, N.Y. did not control bulking. Insufficient information exists on the raCOD loading and nitrate availability in the selector to determine what caused selector failure at these facilities. *N. limicola* was the predominant filamentous bacteria in both cases, which suggests that *N. limicola* is not controlled with anoxic selectors, a finding consistent with Gabb (1988) who grew *N. limicola* in batch-fed systems over many MCRTs.

Table 2-6. Summary of Full-Scale Anoxic Selector Operating and Performance Conditions.

Plant	Beloit, WI	Green Bay, WI		Landis, NJ	Tri-City, OR
		South Plant	North Plant		
Reference	Marten and Daigger, 1997				
Type of Wastewater	Domestic, industrial (food)	Domestic, industrial (paper)		Domestic, industrial (food)	Domestic, industrial (dairy)
Flow, m^3/d	30,100	186,000		31,000	51,100
BOD_5, mg/L	380	180		300	130
Number of Selector Stages	1	1	1	3	1
Selector Volume as Percent of Total Reactor Volume, %	12	8	9	32	20
Aerobic HRT, h	19.8	11.5	12.1	18.7	3.2
Flow Conditions	--	PFR	PFR	PFR	--
Type of Aeration	--	--	--	Surface aeration	--
Temperature, °C	12-18(18-26)	--	--	--	12-18(18-22)
Aerobic MCRT, d	10-12(8-12)	8-11(6-9)	8-11(6-10)	14-33	5.5 (6.8)
Aeration Basin F/M, kg BOD_5/(kgMLSS·d)	--	--	--	--	--
Selector ICZ F/M, kg BOD_5/(kgMLSS·d)	0.7-1.0(0.8-1.2)	1.0-1.3(1.1-1.6)	1.2-1.9(1.4-2.2)	0.63	0.5-1.2(0.7-1.3)
Selector COD Reduction, %	--	--	--	--	--
SVI without Selector, mL/g	--	--	--	--	--
SVI with Selector, mL/g	80-110(70-110)	--	--	75-280	70-156(85-169)
Bulking (B)/Foaming (F) without Selector	--	--	--	--	--
Bulking (B)/Foaming (F), with Selector	--	--	--	--	--
Dominant Filaments without Selector	--	--	--	--	--
Dominant Filaments with Selector	*Nostocoida limicola II*	*N. limicola II*	*N. limicola II*	Type 0041	--
Effect of Selector Installation	Unsteady	Not significant	Not significant	Not significant	Good

Table 2-6. Summary of Full-Scale Anoxic Selector Operating and Performance Conditions (cont'd).

Plant	23rd Ave. Plant, Phoenix, AZ	Area Nolana, Italy	Hardenberg, The Netherlands	Plant A, Italy	Plant B, Italy	Plant C, Italy	Plant D, Italy
Reference	Albertson and Hendricks, 1992	Guida et al., 2002	Kruit et al., 2002	Davoli et al., 2002			
Type of Wastewater	Primarily domestic	Domestic, industrial	--	--	--	--	--
Flow, m^3/d	139,513	81,000	11,385	4,563	3,388	596	587
BOD_5, mg/L	200	147 (sCOD)					
Number of Selector Stages	4	1	1	1	1	1	1
Selector Volume as Percent of Total Reactor Volume, %	17-25	4	5.9	5	3	8	5
Aerobic HRT, h	4.6	8.6	5.5	4.9	5.3	16.1	12.3
Flow Conditions	PFR (stages)	--	--	PFR	CMAS	PFR	PFR
Type of Aeration	Ceramic diffuser	--	--	--	--	--	--
Temperature, °C	--	--	--	--	--	--	--
Aerobic MCRT, d	--	--	--	--	--	--	--
Aeration Basin F/M, kg BOD_5/(kgMLSS·d)	--	--	--	--	--	--	--
Selector ICZ F/M, kg BOD_5/(kgMLSS·d)	8	--	--	3.88(1.4-11.2)	4.6(0.9-14.8)	0.82(0.19-2.4)	1.84(0.3-8.5)
Selector COD Reduction, %		40	--	34	32	40	22
SVI without Selector, mL/g	275	800	--	103 (~170)	120 (~191)	210 (~263)	120 (~222)
SVI with Selector, mL/g	100	49	60-120	56(29-117)	94(36-219)	108(59-167)	118(39-168)
Bulking (B)/Foaming (F) without Selector	B/F	--	--	B/F seasonal	B occasional	B continual	B/F seasonal
Bulking (B)/Foaming (F), with Selector	--	--	--	None	B occasional	None	F seasonal
Dominant Filaments without Selector	Nocardioforms and unidentified species	Type 021N, *Thiothrix*	--	*M. parvicella*	*M. parvicella*, *N. limicola*, Type 0914	*M. parvicella*, Type 0041	*Nocardia*, *M. parvicella*, Type 0041
Dominant Filaments with Selector	--	No	--	*M. parvicella*, Type 0041, Type, 0675	*M. parvicella*, Type 0914	*M. parvicella*, Type 0914	*Nocardia*, Type 0041, Type 0675
Effect of Selector Installation	Good	Good	Good	Positive	Not significant	Positive	None

Table 2-6. Summary of Full-Scale Anoxic Selector Operating and Performance Conditions (cont'd).

Plant	Pulp Mill, Denmark[a]			Mancasale, Italy		
	Phase A	Phase B	Phase C	Test 1	Test 2	Control
Reference	Andreasen et al., 1999			Madoni and Davoli, 1997		
Type of Wastewater	Pulp mill	Pulp mill	Pulp mill	Domestic, industrial	Domestic, industrial	Domestic, industrial
Flow, m^3/d	--	--	--	7,200	7,200	7,200
BOD_5, mg/L	--	--	--	--	--	--
Number of Selector Stages	1	1	1	1	1	0
Selector Volume as Percent of Total Reactor Volume, %	--	--	--	12	5	--
Aerobic HRT, h	--	--	--	5.67	6.33	4.6 (4.1)
Flow Conditions	--	--	--	--	--	--
Type of Aeration	--	--	--	--	--	--
Temperature, °C	--	--	--	18.3	19.5	18.9 (20.5)
Aerobic MCRT, d	--	--	--	8.8	8.4	9.4 (7.4)
Aeration Basin F/M, kg BOD_5/(kgMLSS·d)	--	--	--	0.14	0.28	0.3
Selector ICZ F/M, kg BOD_5/(kgMLSS·d)	10.2[b]	15.8[b]	29.1[b]	1.8	11.48	--
Selector COD Reduction, %	44	8	26	50	--	--
SVI without Selector, mL/g	>400	>400	>400	--	--	100-400
SVI with Selector, mL/g	43	39	>150	--	50-150	--
Bulking (B)/Foaming (F) without Selector	B	B	B	--	--	F
Bulking (B)/Foaming (F), with Selector	--	--	B	--	--	
Dominant Filaments without Selector	Type 021N	Type 021N	Type 021N	--	--	Nocardioforms
Dominant Filaments with Selector			Unknown species	*M. parvicella*	--	--
Effect of Selector Installation	Good	Good	Poor	Not significant	Good	--

Notes: a. Additional nitrate was supplied during anoxic selector investigation.
b. Data converted from kg COD/kg MLVSS to kg BOD_5/kg MLSS basis according to MLVSS/MLSS = 0.85 and BOD_5/COD = 0.6.

2.8.3 Anaerobic Selectors

Installation of an anaerobic selector improved sludge settleability and bulking control at nine of 11 facilities listed (Table 2-7); however, an SVI ≥150 mL/g was reported at four of these nine plants. The inability to develop and sustain a significant PAO population was the primary cause of selector failure at the other facilities (Mill B, Regional WWTP Pilot Plant 2).

Marshall and Richard (2000) reported the poor performance of an anaerobic selector treating wastewater from a pulp and paper mill. A four-stage selector with an overall HRT of 0.75 h and an ICZ F/M of 12 kg BOD_5/(kg MLSS·d) produced an SVI that ranged from 100–450 mL/g. The dominant filament types were *Thiothrix* and *N. limicola III*. Marshall and Richard (2000) suggested that the selector failed to control sludge bulking due to influent wastewater septicity and/or phosphorus deficiency.

Bortone et al. (1995) reported that a single-stage anaerobic selector with an HRT of 0.7 h controlled SVIs to <150 mL/g. The selector volume was 8% of the total system volume and provided an ICZ F/M of 2.0 kg COD/(kg MLVSS·d). Bortone et al. also reported that the selector removed approximately 40% of the total COD.

2.9 Control of Important Filamentous Organisms

This section presents a summary of current knowledge related to control of the most common and problematic filamentous organisms in activated sludge systems—*M. parvicella*, *Thiothrix*, and Type 021N.

2.9.1 Control of *Microthrix parvicella*

M. parvicella has the ability to proliferate under a wide variety of environmental conditions as long as LCFA substrates are present. Based on a review of available literature, the following general statements can be made regarding control of *M. parvicella*:

- Unaerated selectors do not provide effective control. Aerobic selectors may be effective if low DO concentration zones are eliminated.
- Avoid low DO concentrations or intermittent aeration in the main aeration zone, as well as low DO aerobic selectors;
- Reduce MCRT to as low a level as possible while meeting wastewater treatment objectives;
- Design systems with pre-denitrification zones or alternating nitrification and denitrification conditions rather than simultaneous nitrification-denitrification, which often encourages low DO conditions;
- Achieve rapid and complete nitrification, when possible, to maintain low ammonia concentrations;
- Design systems with substrate gradients, such as those found in staged selectors or plug-flow patterns, to facilitate adsorption of LCFA into the sludge flocs;
- Design BNR systems with strict anaerobic or anoxic conditions in the selector stages;
- Remove *M. parvicella* foam as completely as possible and avoid recycling it back through the plant;
- Use RAS chlorination when necessary.

Table 2-7. Summary of Full-Scale Anaerobic Selector Operating and Performance Conditions.

Plant	Fayetteville, AR	Pulp and Paper Mill Plants, USA Mill B	Mill D	Mill G	Seveso WWTP, Italy
Reference	Daigger and Nicholson, 1990	Marshall and Richard, 2000			Bortone et al., 1995
Type of Wastewater	Domestic, industrial	*Thermo-mechanical* pulp (TMP)	Ground wood, semi-chemical mechanical pulp (SCMP)	TMP, de-inking	Textile
Flow, m^3/d	64,000	--	--	--	31,200
Number of Selector Stages	6	4	4	4	1
Selector Volume as Percent of Total Bioreactor Volume, %	17.2	--	--	--	8.00
Aerobic HRT, h	9	--	--	--	8.50
Type of Aeration	Surface mechanical aeration	--	--	--	--
EBPR Performance	Good	--	--	--	--
Temperature, °C	--	--	--	--	15-30
Aerobic MCRT, d	--	--	--	--	12.5
Aeration Basin F/M, kg BOD_5/(kgMLSS·d)	--	--	--	0.15-0.2(0.21-0.25)	--
Selector ICZ F/M, kg BOD_5/(kgMLSS·d)	3.74[a]	48	40-56	48-56	1.41
Selector COD Reduction, %	--	--	--	--	40
SVI without Selector, mL/g	--	--	150-200	--	--
SVI with Selector, mL/g	90	100-450	80-220	80-300	150
Bulking (B)/Foaming (F) without Selector	--	B	--	--	--
Bulking (B)/Foaming (F), with Selector	Sporadic F	B	--	--	--
Dominant Filaments without Selector	--	--	--	--	*Nocardia spp., M. parvicella*
Dominant Filaments with Selector	*Nocardia* during rainfall periods	*Thiothrix I, Nostocoida Limicola III*	*H. hydrossis, Thiothrix I,* Type 0914, Type 0411	Type 0675, Type 1851	*Nocardia spp.*
Effect of Selector Installation	Good	None	Initially poor, then positive	Good, following adjustments	Good settleability, though SVI = 150 mL/g.

Table 2-7. Summary of Full-Scale Anaerobic Selector Operating and Performance Conditions (cont'd).

Plant	Papendrecht, The Netherlands	Alphen, The Netherlands	Groesbeek, The Netherlands	Regional WWTP, CA		
				Full-Scale	Pilot Plant 1	Pilot Plant 2
Reference		Kruit et al., 2002			Fainsod et al., 1999	
Type of Wastewater	Two-Stage Phoredox	Three-Stage Phoredox	Four-Stage Phoredox	--	--	--
Flow, m^3/d	9,540	21,600	4,914	321,000	--	--
Number of Selector Stages	3	4	1	1	4	4
Selector Volume as Percent of Total Bioreactor Volume, %	3.4	5.3	2.8	--	20	20
Aerobic HRT, h	11.8	6.4	10.9	1.76-3.36	3.33	2
Type of Aeration	--	--	--	Pure oxygen		
EBPR Performance	--	--	--		Good	Bad
Temperature, °C	--	--	--	20 (17.5-21.7)	24	24
Aerobic MCRT, d	--	--	--	2.1-4.6	--	--
Aeration Basin F/M, kg BOD_5/(kgMLSS·d)	--	--	--	--	--	--
Selector ICZ F/M, kg BOD_5/(kgMLSS·d)	4.8	11.6	2.6	--	--	--
Selector COD Reduction, %	--	--	--	--	--	--
SVI without Selector, mL/g	--	--	--	--	--	--
SVI with Selector, mL/g	80-170	85-140	80-130	--	--	--
Bulking (B)/Foaming (F) without Selector	--	--	--	--	--	--
Bulking (B)/Foaming (F), with Selector	--	--	--	F	--	--
Dominant Filaments without Selector	--	--	--	Nocardia	*S. natans*, *Thiothrix* sp., Type 021N	*S. natans*, *Thiothrix* sp., Type 021N
Dominant Filaments with Selector	*M. parvicella*	--	*M. parvicella*	--	--	*Thiothrix* sp., Type 021N
Effect of Selector Installation	Good	Good	Good	Nocardioforms not completely eliminated	Eliminated Nocardioforms	None

2.9.2 Control of *Thiothrix*

Since *Thiothrix* thrives on raCOD and uses sulfide oxidation as an energy source, it should be controlled by rapidly removing raCOD using a selector and maintaining a high DO environment in the aeration basin. A review of available literature indicates *Thiothrix* can be controlled by the following:

- Prevent operation under low DO conditions, which may cause septicity;
- Avoid dissolved (soluble) reduced sulfur compounds in the influent feed wastewater;
- Design a sufficiently high ICZ F/M in aerobic selectors to allow the selector to remove as much raCOD as possible;
- Develop a high population of functioning organisms in anoxic (nitrate-reducing organisms) and anaerobic (PAOs) selectors;
- Avoid long retention times in primary clarifiers or stabilization tanks to limit septicity or additional raCOD conversion; avoid recycling septic sludge treatment sidestreams.

2.9.3 Control of Type 021N

Type 021N proliferates primarily in systems with low F/M [<0.2 kg BOD_5/(kg MLSS·d)], high influent raCOD, septic influent, or nutrient deficiency. Bench-scale pure culture experiments have determined that Type 021N has a high affinity for glucose, acetate, and lactate (raCOD compounds) and a competitive growth advantage over floc-forming organisms under nutrient deficient conditions. Control of Type 021N should be centered on preventing these conditions.

2.10 Summary and Conclusions

The objectives of this literature review were to identify operating and metabolic conditions that promote the growth of different filament types, summarize the design and performance of selector applications at pilot- and full-scale facilities, and identify preferred design loadings and configurations.

A number of factors have been identified that promote filamentous bulking in aerobic activated sludge systems, including uptake of raCOD by filaments under aerobic conditions in the main aeration zone, the presence of sulfide compounds, low F/M, low or high wastewater temperatures, uptake of saCOD (LCFAs) instead of raCOD by some filaments types, nutrient deficiency, and low DO concentrations.

The most well-known method for controlling filamentous bulking is the removal of raCOD in a selector zone installed prior to the main aeration zone. In aerobic selector systems, raCOD is removed primarily by floc-forming organisms that have a kinetic advantage over filamentous organisms. This mechanism relies on staged selector designs with a sufficiently high F/M and a long enough HRT to remove virtually all of the raCOD prior to entering the main aeration zone. The literature suggests that the floc-formers have a kinetic advantage in aerobic staged selectors due to a higher uptake rate and storage capacity for raCOD. This mechanism may also be important in anoxic selectors, which suggests that staging anoxic selectors may improve bulking control. However, since floc-formers have a substantially higher denitirification rate relative to filamentous organisms, staging in anoxic selector systems is not as critical as in aerobic selectors. Substrate removal kinetics may also be affected by the main aeration zone design and operating conditions.

Both anoxic and anaerobic selector systems promote raCOD removal by providing floc-formers with a metabolic advantage over filamentous organisms. In anoxic selectors, most filaments are generally unable to use nitrate or nitrite as electron acceptors for raCOD utilization. However, some filaments can proliferate in anoxic systems, which could suggest that providing a kinetic advantage (staged systems) similar to aerobic selectors may be beneficial. A staged anoxic selector may be more favorable for selecting organisms with high substrate storage ability and more rapid uptake of raCOD compared to a single-stage selector. In addition, the staged anoxic selector design could promote an anoxic/anaerobic environment due to the high loading within the first one or two stages. In EBPR systems, the feed-starve cycle present in the anaerobic-aerobic process configuration creates a metabolic advantage for the floc-forming PAOs and/or GAOs, while filamentous organisms do not have the required metabolic capabilities. Staging anaerobic selectors may improve substrate uptake kinetics and provide capacity for handling variable process loading conditions while maintaining a smaller selector volume. Staging anoxic or anaerobic selectors, however, has not yet been demonstrated to consistently improve settleability control.

Pilot- and full-scale design and operating conditions for aerobic, anoxic, and anaerobic selector installations are summarized in Tables 2-5, 2-6, and 2-7, respectively. The range of SVI values achieved by both aerobic and anoxic selectors was broad, with some installations reporting no significant improvement in bulking control. The results indicated that anaerobic selectors generally produced lower SVIs than anoxic selectors.

Typical selector design guidelines are presented in Table 2-4. Major recommended design considerations include staged-selector configurations for aerobic, anoxic, and anaerobic selectors, and limiting process loading values to the ICZ. ICZ values recommended were in the range of 6–12 kg COD/(kg MLSS·d). A review of full- and pilot-scale selector applications and evaluations showed that in many cases single-stage selectors have been used for anoxic and anaerobic selector designs. Staged designs were used quite often for aerobic selectors, but the ICZ values were generally well below the values recommended by Albertson (2005).

Evaluation of pilot- and full-scale applications showed that selector installation alone may not ensure successful SVI control. For example, low DO concentrations or high F/M loading in the main aeration zone may cause poor SVI control. Additional factors influencing poor settleability control are insufficient nitrate return to the anoxic selector zone and nutrient deficiency in industrial wastewaters.

M. parvicella is one of the most commonly occurring filamentous organisms in many parts of the world; thus considerable effort has been made to understand conditions favoring its growth and control methods in activated sludge systems. *M. parvicella* can thrive in plants with a long MCRT, low temperature, LCFAs, and intermittent or local zero-to-low DO concentrations. *M. parvicella* is best controlled by the use of shorter MCRTs and aeration basins with consistently high DO concentrations. There is some evidence that plug-flow or staged aeration basins may also select against *M. parvicella*. An aerobic selector may, at best, provide moderate control of *M. parvicella* by reducing low DO zones in the main aeration zone. Anoxic and anaerobic selectors do not control *M. parvicella*. *M. parvicella* has the ability to take up LCFAs and produce intracellular lipid storage products under both anoxic and anaerobic conditions. It is interesting to note that some plants reported an increase in *M. parvicella* after converting a completely oxic activated sludge system without a selector to an activated sludge system with an anaerobic selector.

The effect on *M. parvicella* growth of staging in aerobic systems has not been adequately addressed to date. Further research demonstrating that selectors do not control *M. parvicella* may not yield additional benefits. At present, the only specific approach (excluding chemical addition) for controlling *M. parvicella* in BNR systems appears to be using a sufficiently low MCRT and maintaining sufficient aeration to avoid any low DO concentration zones or periods in the main aeration basin.

Type 021N commonly causes bulking episodes. It has growth advantages in low F/M systems, requires raCOD substrates, and is favored by septic and nutrient-deficient conditions. Effective control of Type 021N has been achieved when selector conditions allow floc-forming organisms to remove raCOD prior to the main aeration zone.

The following is a summary of the main literature review conclusions:

- A combined survey of 270 U.S. facilities (Jenkins et al. 2004) indicated that the most common filament types were (in order of frequency of occurrence) Type 1701, Type 021N, and *Thiothrix*, while a survey of 33 long MCRT BNR plants in South Africa (Blackbeard et al., 1987) found Type 0092, Type 0675, Type 0041, *M. parvicella*, and Type 0914 to be most common.
- Aerobic selectors promote kinetic conditions favoring preferential substrate uptake and storage by floc-formers over filamentous organisms. Anoxic selectors create a metabolic advantage for floc-formers, since most filament types are unable to denitrify (use nitrate as an electron acceptor) or have relatively low denitrification rates. Similarly, the feed-starve cycle employed in anaerobic selectors allows metabolic selection of floc-forming PAOs or GAOs over filamentous organisms.
- Selectors will be most successful in situations where the target filaments use raCOD as substrates. Selectors may fail if the target filament uses saCOD or sulfide or is favored by low pH or nutrient deficient conditions.
- Some filament types, such as *M. parvicella*, use saCOD (LCFAs) for substrate and will proliferate in selector systems under the following conditions: zero or low DO concentration, long MCRT, and low temperature.
- A review of pilot- and full-scale selector design and operating data showed that a wide range of SVI control was achieved, with some installations reporting no significant improvement in bulking control. Single-stage designs have been employed for anoxic and anaerobic selectors, while most aerobic selectors include a staged-design.

The following is a summary of general selector design guidelines found in the literature:

- Substrate Removal – The sCOD leaving the selector should be <60 mg/L (Shao and Jenkins, 1989) and the raCOD should be virtually absent. The selector should remove 80% of the removable COD (Chudoba and Wanner, 1987).
- Selector Staging and Configuration – All three selector types (aerobic, anoxic, anaerobic) should be designed with at least three stages, sized at 25%, 25%, and 50% of the total selector volume, respectively (Jenkins et al., 2004). A staged-selector arrangement is necessary to create an F/M loading gradient (Albertson, 2005).
- Aerobic Selectors – Aerobic selectors should be staged to provide proper kinetic conditions favoring rapid substrate uptake and storage by floc-formers over filaments. Jenkins et al. (2004) recommended a three-stage design, sized at 25%, 25%, and 50%

of the total selector volume with first stage and total F/M loadings of 12 kg COD/(kg MLSS·d) and 3 kg COD/(kg MLSS·d), respectively.

- Anoxic Selectors – In single-stage arrangements, the selector F/M should be ≤1 kg BOD_5/(kg MLSS·d) for temperatures ≤18ºC and ≤1.5 kg BOD_5/(kg MLSS·d) for temperatures >18ºC, while the anoxic MCRT should be at 1-2 d (Marten and Daigger, 1997). Grady et al. (1999) recommended an anoxic MCRT of 1.0 d at temperatures >20ºC and 1.5 d at temperatures <17ºC. Jenkins et al. (2004) recommended a three-stage design, sized at 25%, 25%, and 50% of the total selector volume with first stage and total F/M loadings of 6 kg COD/(kg MLSS·d) and 1.5 kg COD/(kg MLSS·d), respectively.
- Anaerobic Selectors – A three-stage selector with a total selector HRT of 0.75–2.0 h is recommended (Jenkins et al., 2004).

The following information was highlighted after review of pilot- and full-scale selector systems:

- Staged aerobic selectors yielded significantly lower SVIs compared to single-stage selector systems.
- High SVIs occurred at facilities with high ICZ F/M loadings [~15–20 kg BOD_5/(kg MLSS·d)].
- At recommended ICZ F/Ms, 80%–90% of the COD removed is converted into intracellular storage products. Some oxygen transfer (about 20% of raCOD applied) is needed to provide energy for cellular storage mechanisms.
- Although control of *M. parvicella* was demonstrated in some aerobic selector installations, there was no evidence of consistent and reliable control.
- Anoxic selectors do not effectively control *M. parvicella* growth.
- Single-stage anoxic selector designs yielded variable settling results.
- Since most of the selector ICZ F/Ms were relatively low, the effect of staging on SVI performance was not conclusive.
- Anoxic selectors were not effective at high aeration F/M values [~1.0 kg BOD_5/(kg MLSS·d)].
- Anaerobic selectors appeared to provide lower SVIs than either anoxic or aerobic selectors.
- Anaerobic selectors may provide conditions more favorable for *M. parvicella* growth.

CHAPTER 3.0

LABORATORY INVESTIGATION SUMMARY

3.1 Introduction

As part of this study, a laboratory investigation was conducted to examine the role of feed raCOD (sugars, SCFAs) in the control of filamentous organisms with an aerobic selector. The literature provided evidence that some filamentous bacteria such as *Thiothrix* spp. and Type 021N may be favored by raCOD, while other filamentous bacteria—primarily the "low F/M" filaments such as *M. parvicella* and Type 0092—may be able to use saCOD such as LCFAs. This may explain why low F/M filaments are not usually controlled by a selector. The objective of the laboratory investigation was to grow both an raCOD[1] and an saCOD filament in four reactor systems and to determine whether specific process modifications would control the raCOD filament and not the saCOD filament, including addition of a three-stage aerobic selector to one unit, removal of feed raCOD from the second unit, operation of the third unit as a single-stage CSTR (control), and addition of a four-stage aerobic selector to the fourth unit.

This laboratory investigation was performed at the University of Washington, Seattle by Professor H. David Stensel and Mr. Gang Xin; a summary is presented in this section.

3.2 Materials and Methods

On Day 1, four 3-L completely mixed bench-scale activated sludge units (R1, R2, R3, and R4) were started with activated sludge seed containing significant *M. parvicella* and *Thiothrix* spp. filaments. Following an initial operating period, a three-stage aerobic selector was installed on R1, raCOD was removed from the feed to R2, no changes were made to R3, and a four-stage aerobic selector was added to R4. The bench-scale units were operated at an MCRT of 20 d, a temperature of 12°–15°C, and fed air intermittently (DO between 0–2 mg/L). The systems were fed a synthetic wastewater high in Tween 80 (water soluble oleic acid ester of sorbitol) to promote *M. parvicella* growth and acetate to promote growth of an raCOD filament, such as *Thiothrix* spp. The feed rate to the units was approximately 12 L/d, providing a 6-h detention time in each unit. The reactor operating conditions are summarized in Table 3-1.

Table 3-1. Summary of Bench-Scale Reactor Operating Conditions.

Reactor No.	Description	Operating Conditions (all reactors) Wastewater Feed	MCRT(d)	Temp. (°C)	Air Feed
1	Three-stage aerobic selector	Synthetic, high in LCFAs (oleic acid) and raCOD (acetate)	20	12–15	Intermittent, DO between 0–2 mg/L
2	raCOD removal from feed				
3	single-stage CSTR				
4	Four-stage aerobic selector				

On Day 70, a three-stage selector was installed upstream of R1, and all the raCOD constituents of the synthetic sewage fed to R2 were removed and replaced with the same COD equivalent of saCOD substrates. R1, R3, and R4 continued to receive the synthetic sewage with raCOD included. On Day 99, a four-stage selector was installed upstream of R4. The three-stage

[1] Refer to raCOD discussion on Page 1-3 in Chapter 1.0.

R1 selector consisted of a 200-mL first stage, a 200-mL second stage, and a 400-mL third stage. The four-stage R4 selector consisted of a 140-mL first stage, a 140-mL second stage, a 280-mL third stage, and a 560-mL fourth stage.

Oxygen uptake rate (OUR) batch tests were conducted periodically by adding either acetate (raCOD) or Tween 80 (saCOD) to mixed liquor samples from each of the laboratory units. The OUR test required aerating a 500-mL sample of mixed liquor for 1 h to obtain an endogenous respiration condition, adding 850 mg/L of sodium acetate (400 mg COD/L) or 415/830 mg COD/L of Tween 80, and measuring the decrease in DO concentration (from 6 to 2 mg/L) following repeated aeration cycles.

The DSVI test was used exclusively throughout the experiment to evaluate sludge settleability under various operating conditions.

3.3 Results and Discussion

3.3.1 Diluted Sludge Volume Index

Figure 3-1 shows the DSVI variation over time for all four bench-scale units. The DSVI was already trending down when the three-stage selector was added to R1 on Day 70, but the DSVI continued to drop from 182 mL/g on Day 70 to 107 mL/g on Day 84. Then, the R1 DSVI slowly increased back to 179 mL/g on Day 108. Initially, the R2 DSVI dropped slowly from 226 mL/g on Day 71 (the day after raCOD was removed from the R2 feed) to 170 mL/g on Day 100. Then, the R2 DSVI slowly increased again. The R3 DSVI fluctuated between 150 mL/g and 100 mL/g from Day 1 to Day 93. On Day 108 the R3 DSVI increased to 207 mL/g and increased sharply to 547 mL/g on Day 115. *Thiothrix* spp. was the dominant filament and was microscopically observed at "excessive" levels in R3. The R4 DSVI initially decreased from 119 mL/g on Day 4 to 59 mL/g on Day 12 and then increased to similar values as in R3 (between 150 mL/g and 100 mL/g from Day 17 to Day 70). The R4 DSVI then increased steadily to as high as 500 mL/g (*Thiothrix* spp. dominant filament) on Day 99, when a four-stage aerobic selector was added to R4. The R4 DSVI then decreased to as low as 136 mL/g on Day 114.

Assuming that all four reactors would have had severe bulking (DSVI ≥500 mL/g) due to *Thiothrix* spp. as in R3, which had no selectors and continued to receive raCOD throughout the experiment, then it is possible to conclude that the modifications made to R1, R2, and R4 were successful in controlling severe bulking due to *Thiothrix* spp. This assumption is supported by the R4 DSVI, which also increased to 500 mL/g before a selector was installed. This suggests that controlling *Thiothrix* spp. filamentous bulking with a selector (R1 and R4) is similar to removing the raCOD from the feed (R2). The R2 DSVI, although low compared to the R3 DSVI, was still higher than the R1 and R4 DSVIs by the end of the experiment, suggesting that aerobic selectors may do more to control bulking than just removing raCOD.

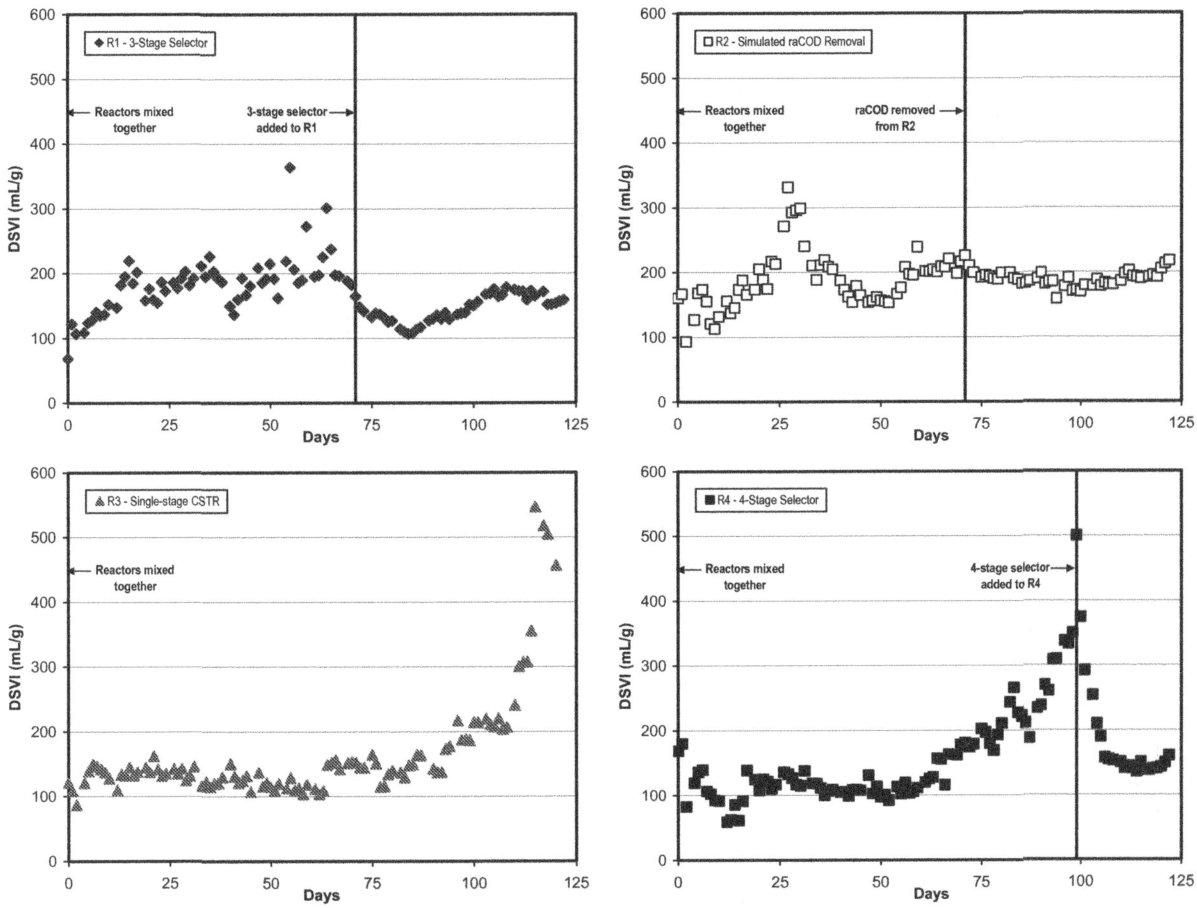

Figure 3-1. Diluted Sludge Volume Index in Four Bench-Scale Reactor Systems.

3.3.2 Microscopic Analysis

A microscopic analysis conducted by Professor David Jenkins on the final mixed liquors of each of the laboratory-scale units is summarized here:

R1: The overall filamentous organism level was "some," which is not sufficient to cause a settling problem. All filamentous organisms were present in small amounts, and those observed were *Thiothrix* I (some with sulfur granules), *M. parvicella*-like filamentous organism, *H. hydrossis*, and Type 1701.

R2: The overall filamentous organism level was "some" which is not sufficient to cause a settling problem. All filamentous organisms were present in small amounts and those observed were *M. parvicella*-like filamentous organism, *H. hydrossis*, and Type 1701.

R3: The overall filamentous organism level was "excessive," which is sufficient to cause severe bulking problems. The major filamentous organism was *Thiothrix* II (some sulfur granules).

R4: The overall filamentous organism level was "some-common," which is not sufficient to cause a settling problem. All filamentous organisms were present in small amounts, and those observed were *Thiothrix* II (some containing sulfur granules), *H. hydrossis*, and an unidentified short Gram and Neisser negative filamentous organism with sausage-shaped cells.

This microscopic analysis indicates that the filament abundance observed for R1, R2, and R4 was similar, so the difference between their DSVIs may be due to non-filamentous factors. If the hypothesis that selectors work by providing an enhanced environment for raCOD floc-formers to outcompete the raCOD filaments for raCOD is true, then the removal of raCOD from the feed to R2 should prevent the growth of these raCOD-type floc-forming bacteria. The raCOD floc-forming bacteria may by themselves enhance floc structure and settleability, and this might explain why the DSVI was higher in R2 compared to R1 and R4.

It was also evident that *M. parvicella* grew poorly or not at all under the conditions of the experiment. This could have been due to the synthetic wastewater composition and/or the operating conditions. Further, *Thiothrix* spp. only dominated in R3, which did not have a selector and continuously received raCOD, leading to the assumption that *Thiothrix* spp. requires raCOD to cause severe bulking, since this was the only difference between R2 and R3.

3.3.3 Batch Testing

Figure 3-2 demonstrates how the R2 OURs and acetate uptake rates changed substantially in batch tests performed before and after raCOD was removed from the R2 daily feed. The R2 mixed liquor batch OUR peaked at approximately 175 mg/(L·min) prior to raCOD removal from its feed, 70 mg/(L·min) 15 days after raCOD was removed, and only about 18 mg/(L·min) 56 days after raCOD was removed. Similarly, acetate uptake from these batch tests was measured at about 10 mg/(L·min) prior to raCOD removal, 2 mg/(L·min) 15 days after raCOD was removed, and about 0.9 mg/(L·min) 56 days after raCOD was removed. This demonstrates a significant population shift in the R2 mixed liquor in response to the removal of raCOD from the R2 feed, with the resulting population unable to take up acetate rapidly.

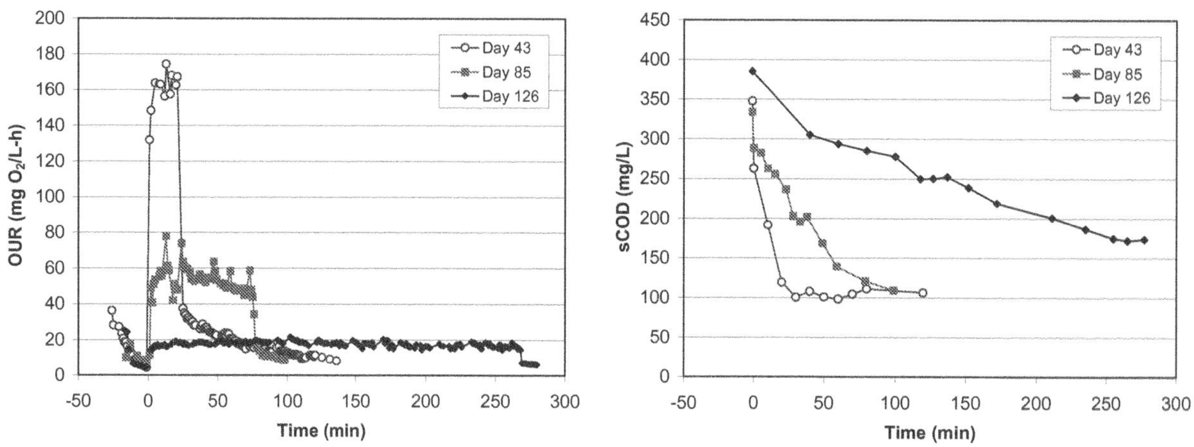

Figure 3-2. Oxygen Uptake Rate (OUR) Profiles and Soluble COD Concentration During OUR Tests with 850 mg/L Sodium Acetate Addition to R2.

WERF

Figure 3-3 compares the OURs and acetate uptake rates from each of the four laboratory units (R1, R2, R3, and R4) around Day 120 of the experiment. This figure demonstrates how both the OUR and acetate uptake rates are significantly lower for R2 than for the other three units. The average peak OUR for R2 is about two–six times lower than R1, R3, and R4, and the acetate uptake rate for R2 is about three–seven times lower than the other three laboratory units. This suggests a significant population difference between the R2 mixed liquor, with a shift away from those organisms that rapidly take up acetate, compared to the other units. These results support the hypothesis that the R2 feed without raCOD may not have supported the growth of raCOD floc-forming bacteria, and the presence of these bacteria may enhance floc structure and settleability.

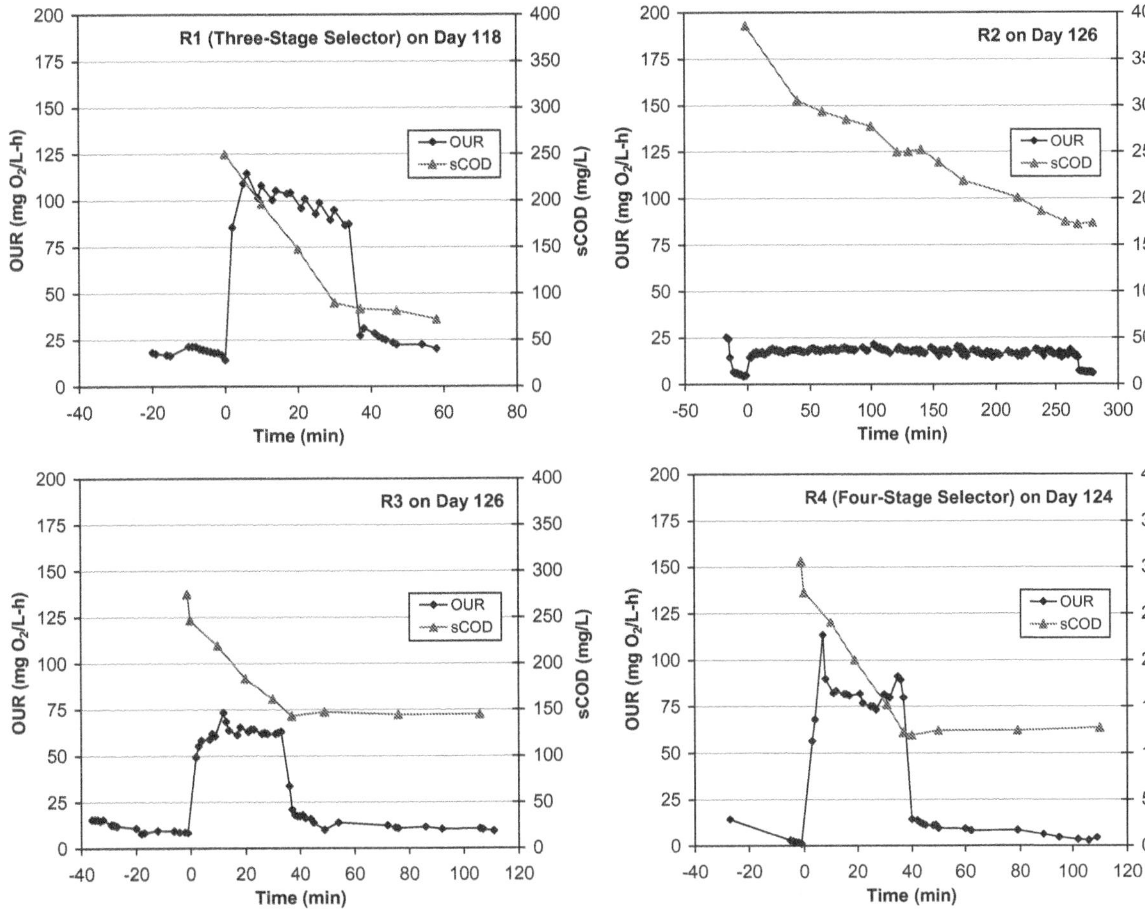

Figure 3-3. Oxygen Uptake Rate (OUR) Profiles and Soluble COD Concentration during OUR Tests with 850 mg/L Sodium Acetate Addition to Four Bench-Scale Reactor Systems.

Figure 3-4 compares the mixed liquor OURs and Tween 80 uptake rates from each of the four laboratory units about 15–20 d after the three-stage aerobic selector was added to R1 and raCOD was removed from the R2 feed. Both R3 and R4 were operated without a selector during this period. The Tween 80 uptake rates were between 0.6–0.9 mg sCOD/(L·min) for all the laboratory units and were closer to each other than the acetate uptake rates. Moreover, the acetate uptake rates were about six–10 times higher than the Tween 80 uptake rates. This suggests that Tween 80, and possibly all LCFAs, may not be adequately removed in a selector and could therefore leak into the main aeration basin in sufficient quantities to support significant filamentous organism growth and cause severe bulking.

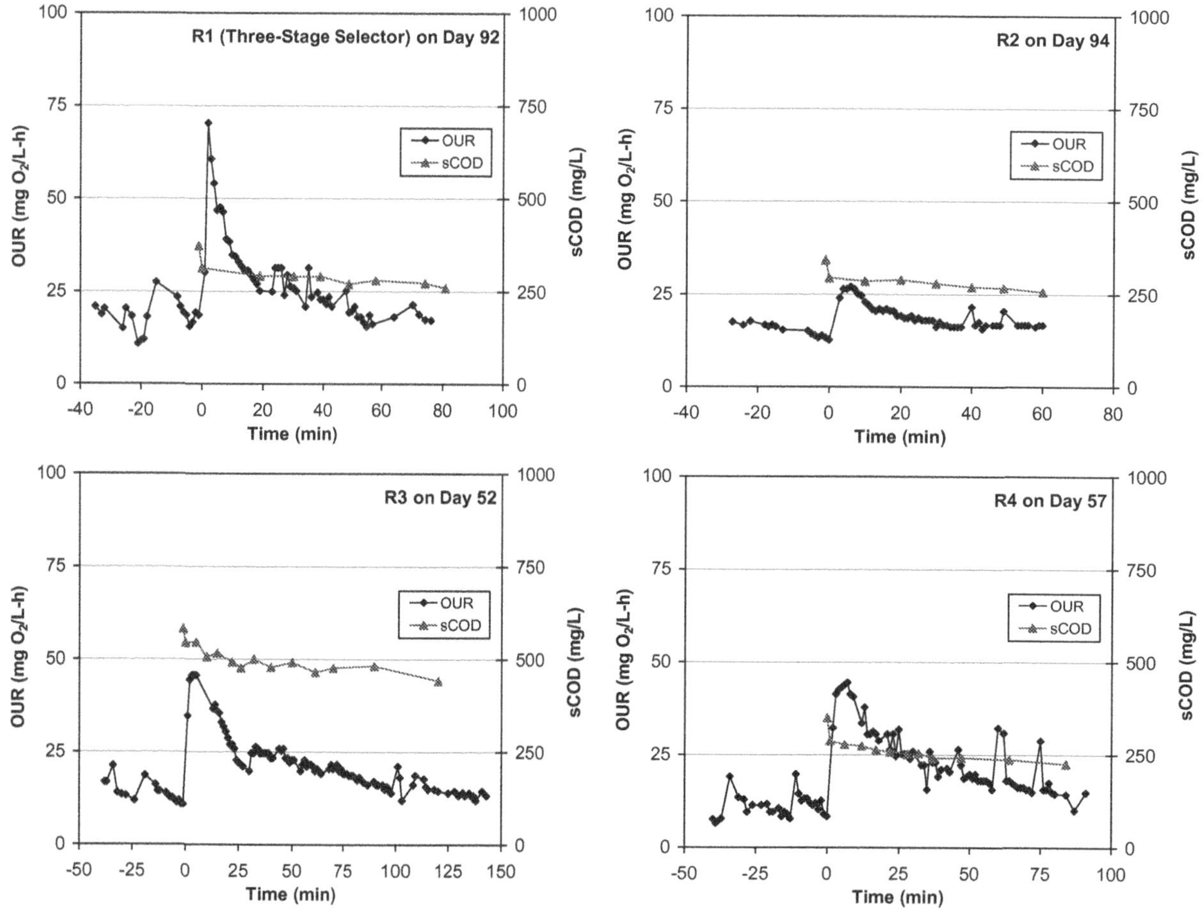

Figure 3-4. Oxygen Uptake Rate (OUR) Profiles and Soluble COD Concentration During OUR Tests with 220 mg/L Tween 80 Addition (440 mg/L for R3) to Four Bench-Scale Reactor Systems.

3.3.4 sCOD Uptake Through the Selectors

Tables 3-2 and 3-3 show the sCOD uptakes through the three- and four-stage aerobic selectors. Most of the sCOD was taken up in the first stage of both selectors, and very little if any sCOD was taken up in the remaining stages. Based on this data, the four-stage selector should perform no better than the three-stage. It is difficult to see this from the data collected in this study, since the DSVI and DSVI trends prior to adding a selector were different for R1 and R4. Nonetheless, Figure 3-1 shows that the DSVI was essentially the same for both units at the end of the experiment, which suggests similar performance for these systems.

Table 3-2. Soluble COD Concentration (mg sCOD/L) Measured Across Three-Stage Selector Reactor (R1).

Date	Influent	Applied	Stage 1	Stage 2	Stage 3	Effluent	RAS
5/10/03	634	402 – 475	335	274	280	248	--
5/18/03	766	442 – 544	311	292	291	226	--
5/25/03	662	398 – 481	288	265	251	222	--
6/1/03	630	424 – 488	299	286	267	246	286
6/14/03	456	289 – 341	163	166	173	135	177
6/23/03	476	303 – 357	215	193	187	166	188

Note: "Applied" sCOD was calculated based on a 0.7 or 1.5 RAS rate.

Table 3-3. Soluble COD Concentration (mg sCOD/L) Measured Across Four-Stage Selector Reactor (R4).

Date	Influent	Applied	Stage 1	Stage 2	Stage 3	Stage 4	Effluent	RAS
6/7/03	601	414 – 472	282	259	267	252	229	289
6/14/03	420	274 – 319	165	165	168	154	135	176
6/23/03	524	329 – 390	210	206	211	200	176	199

Note: "Applied" sCOD was calculated based on a 0.7 or 1.5 RAS rate.

3.4 Conclusions

This experiment generated the following conclusions:

- Severe bulking (DSVI ≥500 mL/g) due to *Thiothrix* spp. was controlled or prevented with three-stage and four-stage aerobic selectors and by removing raCOD from the wastewater fed to an activated sludge process. This suggests that raCOD removed from the main activated sludge aeration basin, either with a selector or by excluding it from a synthetic sewage fed to the activated sludge process, significantly reduces the growth of *Thiothrix* spp.
- Removing raCOD from wastewater fed to activated sludge processes alone may not produce DSVIs as low as activated sludge processes equipped with a well-performing selector. This may be because selectors enhance the growth of raCOD floc-forming bacteria, while activated sludge processes fed wastewaters absent of raCOD do not support the growth of these organisms; and raCOD floc-forming bacteria may enhance activated sludge floc structure and settling on their own.
- Uptake rates of Tween 80 and possibly LCFAs are six–10 times slower than those of acetate. This suggests that even well-performing selectors may not adequately remove LCFAs and could allow them to leak into the main aeration basin where they may be used by filamentous organisms for growth.

CHAPTER 4.0

DETAILED PLANT INVESTIGATIONS

4.1 Introduction

Many of the selector case studies reported in the literature lack critical pieces of information necessary to properly evaluate selector design and performance. Typically, average SVI values are reported without ranges. Many facilities are not able to provide data on filament type and abundance prior to and following selector installation. Moreover, insufficient data is provided to compare selector design parameters between facilities. This lack of information has hindered the development of selector design guidelines.

The goal of the detailed plant investigations task was to collect selector design and operating data from a large number of facilities equipped with aerobic, anoxic, or anaerobic selectors. Based on the information collected, typical selector design parameters were examined to determine whether a correlation existed with settling performance. This section presents the results of a detailed evaluation of full-scale selector installations with a specific focus on selector design and associated SVI control.

4.2 Initial Screening Survey

An initial screening survey of 125 U.S. plants was completed to identify candidate facilities for more detailed study. A total of 85 of the 125 plants were reported to have an aerobic, anoxic, or anaerobic selector in the secondary treatment process; however, only 46 of these facilities reported improved settleability following selector installation, as shown in Figure 4-1. The initial screening survey is provided in Appendix A.

The specific objectives of the initial screening survey were to identify a pool of facilities equipped with selectors of various types, establish contacts at each facility, and gauge the level of interest from each facility in participating in the detailed field investigation. Given the significant amount of additional plant data to be requested and assuming a moderate response rate, the project team decided to carry forward all 85 facilities from the initial screening reporting selector installations.

4.3 Data Collection, Processing, and Verification

4.3.1 Data Collection

Table 4-1 summarizes the information requested from each of the 85 facilities included in the detailed plant investigation. In addition to collecting general plant and process configuration information, the project team requested that each facility provide approximately one year of selector operating and performance data in spreadsheet format. The detailed plant survey is provided in Appendix A.

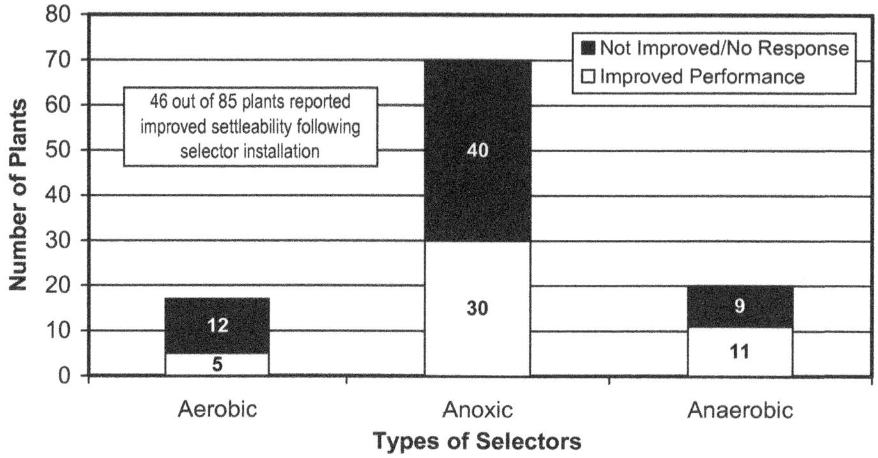

Figure 4-1. Initial Screening Survey Results – Selector Type and Effectiveness.
Note: Some plants reported the presence of multiple selector types.

Table 4-1. Summary of Detailed Plant Investigation Data Requested.

Category	Description
General Information	• Facility name, location, contact • Average, peak flow rate • Industrial contribution, major contributors • Annual wastewater temperature range • Nutrient removal requirements and processes
Selector Configuration	• Selector type (aerobic, anoxic, anaerobic) • Number and volume of selector stages • Mixing type (hydraulic, mechanical, air) • Available process design criteria, technical reports
Aeration Basin Configuration	• Number and volume of aeration stages and basins • Type of aeration system • Internal recycle streams • Approximate dissolved oxygen DO profiles • Location of return activated sludge RAS feed points
Additional Plant Information	• Process schematic • Secondary process operation and maintenance (O&M) manuals • Secondary influent sulfide levels • Oxygen uptake rate data • Soluble BOD or COD exiting the selector zone
Plant Operating Data (One Year)	• Secondary influent – flow, BOD, sBOD, COD, sCOD, TKN, P • Number of aeration basins in-service • WAS, RAS flow and concentration • MLSS, MLVSS • System (excluding clarifier solids), aerated MCRT • F/M • DO • Influent or effluent pH • SVI or DSVI • Filament type and abundance • RAS chlorination periods

4.3.2 Data Processing

This section describes the data calculations and estimates applied to the selector operating data provided by the 44 full-scale facilities.

4.3.2.1 Process Data Calculations

Average values for typical activated sludge process parameters were calculated from the plant operating data provided by each facility in spreadsheet format. Based on the information provided, the project team calculated a number of important selector design and operating parameters, as summarized in Table 4-2. A more detailed discussion of each parameter calculation is included in Appendix C.

Table 4-2. Summary of Detailed Plant Investigation Process Data Calculations.

Parameter	Comments
Selector MCRT (d)	Calculation based on mass of mixed liquor in selector zone only
Contact (or floc) loading (kg BOD_5/kg MLSS)	Ratio of influent BOD mass to mass of solids in the ICZ
Selector ICZ F/M loading [kg BOD_5/(kg MLSS·d)]	F/M calculation based on mass of mixed liquor in selector ICZ only
Nominal Selector HRT (without recycle) (h)	HRT calculation based on volume of selector zone only
Nominal Selector ICZ HRT (without recycle) (h)	HRT calculation based on volume of selector initial contact zone only
Selector HRT (with recycle) (h)	HRT calculation based on volume of selector zone only, includes mixed liquor recycle and RAS flows
Selector ICZ HRT (with recycle) (h)	HRT calculation based on volume of selector ICZ only, includes mixed liquor recycle and RAS flows
Effective Number of Selector Stages	Estimated based on semi-empirical formula (see Section 4.3.2.3)
90th Percentile SVI (mL/g)	
90th Percentile Merkel DSVI (mL/g)	SVI data converted to DSVI using Merkel equation (see Section 4.3.2.2)
Fraction of SVIs greater than 150 mL/g (%)	Represents percent of time SVIs exceed typical control limit

4.3.2.2 Estimating DSVI from SVI Data

An activated sludge settleability index should be independent of the solids concentration in order to be considered meaningful. As stated in Section 1.1, the SVI test result is dependent on solids concentration at SSV_{30} values greater than 300 mL/L. The DSVI test proposed by Stöbbe (1964) overcomes the concentration dependence issue and is the desired settleability index for the purposes of this analysis. Merkel (1971) evaluated a large number of settling test results and developed the following formula to convert a measured SVI value to DSVI:

$$\text{DSVI}, \text{mL/g} = \text{SVI}, \text{mL/g}\left(\frac{300}{\text{SSV}_{30}, \text{mL/L}}\right)^{0.6} \quad (4.3.2.2\text{-}1)$$

where: SVI = sludge volume index (mL/g)
SSV_{30} = 30 min settled sludge volume (mL/L)

SSV_{30} values were back-calculated for each data point when not provided by the plant, according to the following equation:

$$\text{SSV}_{30}, \text{mL/L} = \text{SVI}, \text{mL/g}\left(\frac{\text{MLSS}, \text{mg/L}}{1000\ \text{mg/g}}\right) \quad (4.3.2.2\text{-}2)$$

If the SSV_{30} was ≤300 mL/L, then the SVI qualified as the DSVI and was reported as such. When the SSV_{30} was >300 mL/L, the SVI was converted to the DSVI according to the equations presented above. Figure 4-2 illustrates the effect of applying the Merkel DSVI correction to SVI data.

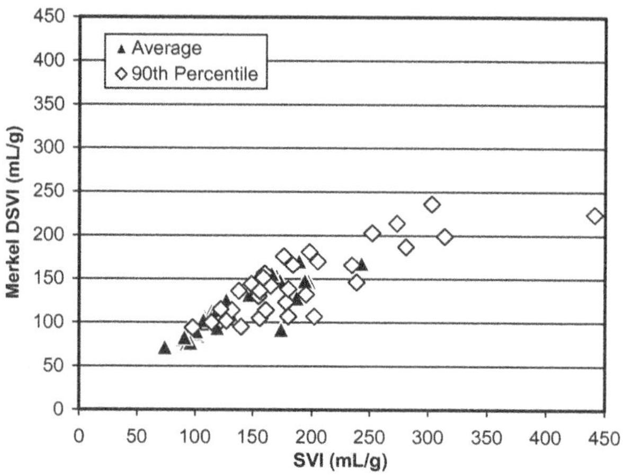

Figure 4-2. Average and 90th Percentile SVI and DSVI Comparison.

4.3.2.3 Estimating Effective Number of Selector Stages

Chudoba et al. (1973) and others showed how SVI and filamentous organism abundance dropped with a decreasing vessel dispersion number (D/uL) in fully aerated activated sludge systems. As D/uL decreases, the reactor more closely approximates plug-flow conditions (negligible dispersion), and when D/uL = ∞ the system is considered a completely mixed (high dispersion) reactor (Levenspiel, 1972). Although a low dispersion number has been shown to be essential for aerated selector performance, Wanner (1994) has shown that anaerobic selectors can eliminate filamentous bulking even when the selector is a single-stage, well-mixed basin (D/uL → ∞). It is unclear, however, whether there are any significant benefits to reducing D/uL in anaerobic or anoxic selectors.

Since D/uL is derived from dye studies, and it was impractical to conduct dye studies on all the full-scale wastewater treatment plants studied in this project, D/uL could not be directly evaluated in this study. To address dispersion, the project team developed a dimensionless, semi-empirical number (N) to approximate the dispersion characteristics of the full-scale selector basins included in this study. N was calculated according to the following equation:

$$N = 6.04 + \log_{10} (V\, L^{1.333}\, n / 1.486\, w^2) \qquad (4.3.2.3\text{-}1)$$

where: V = flow velocity through the selector basin (ft/s)
L = selector basin length (ft)
w = selector basin width (ft)
n = Manning coefficient ($s/m^{0.333}$), dividing by 1.486 converts n from $s/m^{0.333}$ to $s/ft^{0.333}$, and makes N a dimensionless number.

Note: The flow velocity (V) is calculated by summing all of the flows (secondary influent and all recycle flows) to the selector basin and dividing by the basin cross-sectional area (basin width times depth). High V values, however, can distort N to slightly higher than reasonable values. Therefore, V should be limited to that velocity that provides at least a 45-min theoretical selector HRT (total net flow to the basin divided by basin liquid volume), regardless of whether the actual selector HRT is less than 45 min.

Adding 6.04 to the equation provides an approximation of the equivalent number of tanks, compartments, or stages in series for mixing conditions in the basin considered. This number was determined by using the equation on a variety of both theoretical and actual basin conditions. This equation is not intended to replace dye testing, but instead is intended to provide a reasonable approximation for the number of equivalent or effective stages when dye testing is not feasible.

N Equation Derivation Summary

Harleman's (1964) eddy diffusion coefficient (E) equation for estuarine systems was first considered:

$$E = C\,n\,v\,R^{5/6} \qquad (4.3.2.3\text{-}2)$$

where: C = a constant
n = Manning's roughness coefficient
v = flow velocity (ft/s)
R = hydraulic radius (ft)

Since long narrow channels are usually considered to have flow regimes more similar to plug-flow conditions compared to tanks that are more cube-shaped, which are considered to have completely mixed flow regimes, the channel or basin length-to-width ratio was considered to be more applicable to basin dispersion than hydraulic radius. $L^{1.333}/w^2$ replaced $R^{5/6}$ and the constant (C) was removed, but 1.486 was added to provide a dimensionless number in English units. The number range generated from this relationship covered orders of magnitude. To shorten this range, the $\log_{10}$ of this number was used in the equation. Applying this logarithmic relationship to theoretical and actual basins showed that an approximate number of "tanks-in-series" could be obtained if 6.04 was added to the number calculated from the logarithmic relationship.

Examples Using N

Completely Mixed Continuous-Flow Stirred Tank Reactor (CSTR)

A typical CSTR could be modeled as a cubically structured tank with the width, length, and depth all being equal. The flowrate would be such that the HRT = 45 min to comply with the limitations on the flow velocity, V. Assuming that the tank is 50 ft × 50 ft × 50 ft and the flowrate = 46.3 ft^3/s, then V would equal 0.0185 ft/s. For a concrete-lined channel, n = 0.014 was assumed. Then:

$$N = 6.04 + \log_{10}(0.0185 \times 50^{1.333} \times 0.014 / 50^2 \times 1.486)$$

N = 1.15 (or approximately one tank in series)

Long Narrow Channel

For a long narrow channel, assume length = 250 ft, width = 10 ft, and depth = 10 ft. Assuming a 45-min HRT, flow = 9.26 ft^3/s and V = 0.0926 ft/s. The Manning coefficient, n, is again assumed to be 0.014. Then:

$$N = 6.04 + \log_{10}(0.0926 \times 250^{1.333} \times 0.014 / 10^2 \times 1.486)$$

N = 4.18 (or approximately the equivalent of four tanks in series)

East Bay Municipal Utility District

The East Bay Municipal Utility District (EBMUD) activated sludge reactor basins are 46 ft long, 46 ft wide, and 25 ft deep. Combined secondary influent and RAS flows are about 34 ft^3/s on average for each selector-equipped process train, which provides a 26-min HRT. To calculate N, however, the flowrate or velocity through the basin, V, must be no more than that to provide a 45-min HRT. Therefore, the flowrate used in the calculation must be 19.6 ft^3/s, for a V = 0.017 ft/s. Then:

$$N_{EBMUD} = 6.04 + \log_{10} (0.017 \times 46^{1.333} \times 0.014 / 46^2 \times 1.486)$$

$$N_{EBMUD} = 1.18 \text{ (or approximately one tank in series)}$$

Upper Occoquan Sewage Authority

The Upper Occoquan Sewage Authority (UOSA) selector basins are 55 ft long, 11 ft wide, and 15 ft deep. Average combined flow (secondary influent and RAS) to each selector is approximately 12 ft^3/s, which results in a 7.9-min HRT. To comply with the minimum 45-min HRT requirement, 3.4 ft^3/s was used to calculate N, resulting in V = 0.0206 ft/s. Then:

$$N_{UOSA} = 6.04 + \log_{10} (0.0206 \times 55^{1.333} \times 0.014 / 11^2 \times 1.486)$$

$$N_{UOSA} = 2.61 \text{ (or approximately 2.5–3 equivalent tanks in series)}$$

Previously, a dye study was performed at UOSA, and the selector compartments were determined to be the equivalent of three tanks in series. This compares reasonably well to the N-calculated number of tanks in series. The flowrate was much higher in the actual UOSA selectors and might have contributed to the slightly higher number of tanks in series number determined from the dye study compared to the N number. On the other hand, the N number is only intended to approximate what the dye study "measures."

Table 4-3 shows the comparison between the physical number of selector compartments and the equivalent number of selector stages calculated with the N number.

4.3.2.4 Plant Data Analysis

As an initial step, average values for selected selector design and operating values were plotted against 90th percentile SVI and DSVI values for each of the facilities by selector type (aerobic, anoxic, anaerobic). This approach is somewhat limited because each plant is only represented by a single data point, which does not consider how the variation in each parameter may influence variation in SVI or DSVI. Therefore, a single-variable regression analysis, incorporating daily operating data from each facility, was conducted to better evaluate the influence of parameter variation on SVI and DSVI values. Further, the plant data sets were divided into three distinct plant categories—short MCRT with anoxic or anaerobic selectors, short MCRT with aerobic selectors, and long MCRT with selectors—since different filamentous organisms are present in short- versus long-MCRT systems. This categorization also accounts for different substrate uptake and growth mechanisms in aerobic (kinetic) versus anoxic and anaerobic (metabolic and kinetic) selector systems. The methodology used to categorize the plants is explained in the next section.

4.3.2.5 Categorizing Plant Data Sets

Three plant characteristics were used in concert to categorize facilities as either short- or long-MCRT plants (listed in order of importance): 1) type of predominant filamentous organism present, 2) system MCRT (excluding clarifier solids), and 3) nitrogen removal.

1. **Predominant filamentous organisms present.** *S. natans, Thiothrix* sp., Type 1863, Type 021N, and Type 1701 are usually found in short-MCRT plants, whereas *M. parvicella, N. limicola,* Type 0092, Type 0041, Type 0675, Type 0914, and Type 1851 are typically found in long-MCRT plants. The type of filament identified at each plant was the primary criterion in cases where reliable, relevant, and conclusive filament data was available. In cases where the filament data provided was unreliable or if both short- and long-MCRT filaments were identified at equal levels and frequencies of occurrence, the two other criteria (system MCRT and nitrifcation) were used to classify the plant.
2. **System MCRT (excluding clarifier solids).** Plants with a short system MCRT (<4 d) were classified as short-MCRT plants, while plants with a long system MCRT (>10 d) were classified as long-MCRT plants. In cases where the system MCRT was between 4–10 d, the other two criteria were used to classify the plant.
3. **Nitrogen Removal.** Nitrifying plants were categorized as long-MCRT plants, whereas non-nitrifying plants were classified as short-MCRT plants.

The majority of the plants were relatively easy to classify as either a long- or short-MCRT plant based on the above criteria. A small number of plants, however, were more difficult to classify. These plants are discussed individually below.

1. **Yakima WWTP (Yakima, Wash.)** – This plant had an MCRT of 11 d and demonstrated complete nitrification during the study period. However, it was classified as a short-MCRT plant because short-MCRT filaments (*S. natans, Thiothrix* sp., and Type 021N) were identified by plant personnel during bulking epidoses within the study period. According to plant personnel, bulking occurs seasonally due to industrial discharges from pear canneries. The activated sludge influent BOD_5/TSS ratio also increased significantly during the bulking period, suggesting an increase in soluble organic loading.
2. **OMI Plant 6 (South Central US)** – Although this plant had a system MCRT of 12 d and is a nitrifying plant, it was classified as a short-MCRT plant based on frequent observations by plant personnel of Type 021N and *Thiothrix* in the mixed liquor.
3. **Southside WWTP (Tulsa, Okla.)** – The Southside WWTP had an MCRT of 4.8 d (closer to 4 d than 10 d), and full or partial nitrification often occurred during the study period. A mixed liquor sample was obtained from the plant and was found to contain no dominant filaments. However, a number of long-MCRT filaments were observed in the sample (Type 0092, Type 0675, and Type 0041). This plant was therefore placed in the long-MCRT category.
4. **Glendale WWRF (Lakeland, Fla.)** – Although Glendale had an MCRT of 5.3 d (closer to 4 d than 10 d), the plant generally nitrifies and operates a mixed liquor recycle line. In addition, long-MCRT filaments (*N. limicola II and III*) were identified as dominant filaments in the mixed liquor during an analysis performed in 2001 (prior to the study period). This plant was categorized as a long-MCRT plant.
5. **Billings WWTP (Trains 2 and 3) (Billings, Mont.)** – The two Billings WWTP process trains were both categorized as short-MCRT plants because they do not nitrify and have short MCRTs (< 4d). Mixed liquor samples from each train were analyzed for filament characterization as part of the detailed survey. *M. parvicella*

was dominant in both Trains 2 and 3, and Type 1701 was also found to be dominant in Train 3. However, prior filament analyses have regularly identified *M. parvicella*, *Thiothrix* and Type 1701 as dominant filaments, and Type 1863 has appeared when the trains were operated with a short MCRT. The filament information was inconclusive, and the plants were therefore classified based on their short MCRT and lack of nitrification.

6. **OMI Plant 4 (Midwestern US)** – This plant had an average MCRT of 4.8 d and is generally a nitrifying plant. *M. parvicella* was identified as a dominant filament in 2001 (prior to the study period). However, the period of bulking during the study period occurred specifically during a period of low MCRT (3–4 d), and high aeration basin DO was reported, which reduced the likelihood that *M. parvicella* was present at this facility during the study period. In this case, the existing filament data was likely not representative of conditions during the period of higher DSVIs. Therefore, this plant was placed in the short-MCRT category.

4.3.2.6 Interpolating Missing Process Data

Each "plant-day" of operating data was used as an input into the single-variable regression analysis. However, many parameters are measured less frequently than on a daily basis at many plants, or for other reasons are missing data for certain days. Estimated values based on a linearly weighted moving average formula were used in place of missing values where applicable.

Estimated values were calculated to complete missing values for the following parameters: BOD_5, TSS, MLSS, temperature, and aeration basin DO concentration. Data with missing values was handled in time-series data ranges, each of a single parameter (BOD_5 data, for example) from a single plant. For each missing value, an estimated value was calculated by using actual measured data within a specified interval of x days before and after the missing value.

All valid measured data points within this interval were used to calculate a linearly weighted moving average. This method multiplies measured data within the interval by a weighting factor that emphasizes measured data collected shortly before or after the day for which a value is being currently calculated. For an interval of six days, a measured value from one day before or after the missing value was multiplied by six, a measured value from two days before or after the missing value was multiplied by five, and so on, up to measured values from six days before or after the missing value, which were multiplied by 1.

Table 4-3 presents the calculation of an estimated BOD_5 value for October 26, 2002, a day for which no measured BOD_5 value was available. Three measured values from before the current day (184, 181 and 158) and three measured values from after the current day (190, 191 and 166) were used to calculate a BOD_5 value for October 26 as follows:

$$\text{Calc.BOD}_5 = \frac{184\cdot 2 + 181\cdot 3 + 158\cdot 5 + 190\cdot 6 + 191\cdot 4 + 166\cdot 2}{2+3+5+6+4+2} = 179.0 \qquad (4.3.2.6\text{-}1)$$

The same formula was used to calculate an estimated value for each missing data point in the data range where there was at least one measured value before the missing value and one measured value after the missing value within the specified interval.

An interval of six days in each direction (before and after the missing value) was used for all but three data ranges because this interval allowed for estimates of six consecutive missing data points in data ranges with weekly measured data. An interval of six days was also small enough to avoid the excessive "smoothing" of time-series data that occurred when larger time intervals were used.

Table 4-3. Example Calculation for Estimating BOD_5 Value Using Linearly Weighted Moving Average.

Date	BOD_5 (Measured)	Time	Weighting Factor	Weighted Sum	BOD_5 (Calculated)
20-Oct-02	No Data	t = -6	1	--	
21-Oct-02	**184**	**t = -5**	**2**	**368**	
22-Oct-02	**181**	**t = -4**	**3**	**543**	
23-Oct-02	No Data	t = -3	4	--	
24-Oct-02	**158**	**t = -2**	**5**	**790**	
25-Oct-02	No Data	t = -1	6	--	
26-Oct-02	*missing*	*t = 0 (current)*	--	--	*179.0*
27-Oct-02	**190**	**t = 1**	**6**	**1140**	
28-Oct-02	No Data	t = 2	5	--	
29-Oct-02	**191**	**t = 3**	**4**	**764**	
30-Oct-02	No Data	t = 4	3	--	
31-Oct-02	**166**	**t = 5**	**2**	**332**	
01-Nov-02	No Data	t = 6	1	--	
Sum[1]			**22**	**3937**	

Note:
[1] Sum of weighting factors only includes dates on which a measured BOD5 concentration was available.

Figures 4-3 and 4-4 presents graphically the results of the missing value estimation process for two data ranges:

- **BOD_5 data for OMI Plant 4.** This data range contains two–three measured data points per week. Variability of data from day to day is moderate. Calculation of estimated values resulted in some smoothing of data, but individual outliers still affected the calculated values significantly, pulling the trend of calculated values up or down.
- **Wastewater temperature data for Veolia – Plant 1.** This data range contains weekly data, with a period of about two months without data. Estimated values were essentially a linear interpolation of the two surrounding data points. The missing values in the period with no data were not estimated, since these missing values did not meet the requirement of having at least one measured data point before and after each missing value within the specified interval.

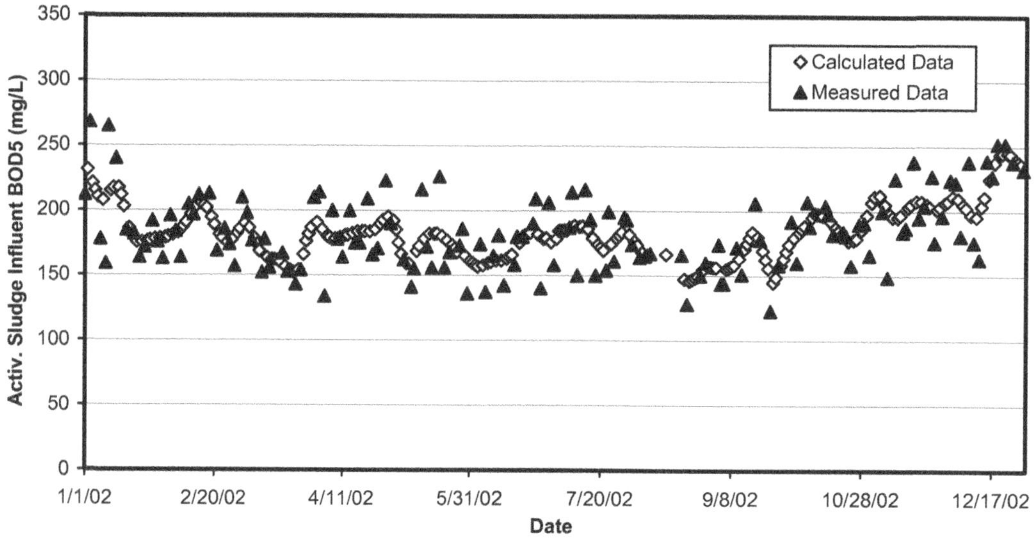

Figure 4-3. Measured and Calculated Activated Sludge Influent BOD_5 Values for OMI Plant No. 4.

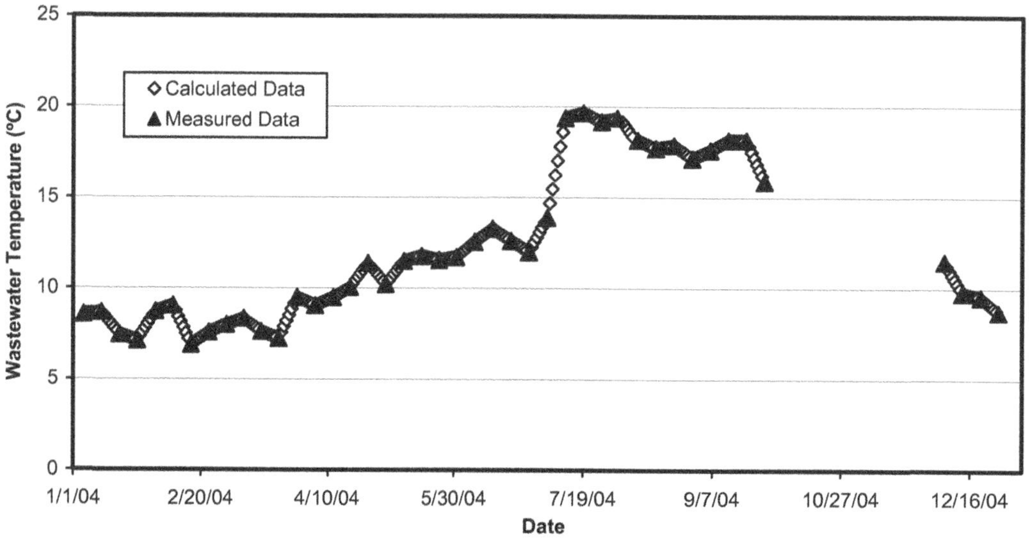

Figure 4-4. Measured and Calculated Wastewater Temperature Values for Veolia Plant No. 1.

Three data ranges contained actual data collected at a frequency of once every two weeks: BOD_5 and TSS data from OMI - Plant 3 and temperature data from Upper Occoquan Sewage Authority. The missing values in these data ranges were estimated using a larger interval of 13 days before and after each missing value.

A summary of the interpolated data for each parameter in each of the three plant categories is presented in Table 4-4.

Table 4-4. Summary of Interpolated Data for Specific Process Variables.

Plant Category	Interpolated Data					
	BOD_5	TSS	MLSS	DSVI	DO	pH
Short MCRT Aerated	39% (404 of 1031)	21% (216 of 1035)	0.4% (4 of 1035)	25% (259 of 1025)	37% (268 of 730)	0.1% (1 of 1012)
Short MCRT Unaerated	37% (1878 of 5125)	31% (1588 of 5149)	9% (488 of 5148)	11% (572 of 5107)	1% (53 of 3682)	9% (441 of 4963)
Long MCRT	40% (3539 of 8793)	31% (2603 of 8466)	9% (806 of 9196)	11% (951 of 8982)	7% (528 of 8082)	11% (927 of 8377)

4.3.3 Data Verification

Given the large amount of information requested from each facility, a significant level of effort was allocated toward verifying the information provided and clarifying missing or outlier data. In many cases, plants were not able to provide key information, such as filament type and abundance, SVI, or secondary process operating data. The project team was successful in collecting and verifying data from 44 of the 85 plants originally included in the detailed plant investigations. The next section focuses on selector performance and operating data from these facilities.

4.4 Results and Discussion

The results of the detailed plant investigations of 44 full-scale facilities equipped with aerobic, anoxic, or anaerobic selectors are summarized in this section. A summary of the detailed plant evaluation data collected is presented in Table 4-5.

4.4.1 Facility Size and Selector Type Distribution

A summary of the facility size, based on influent flow rate, and selector type is presented in Figure 4-5. The detailed plant investigation included five aerobic selector, 27 anoxic selector, and 16 anaerobic selector installations.

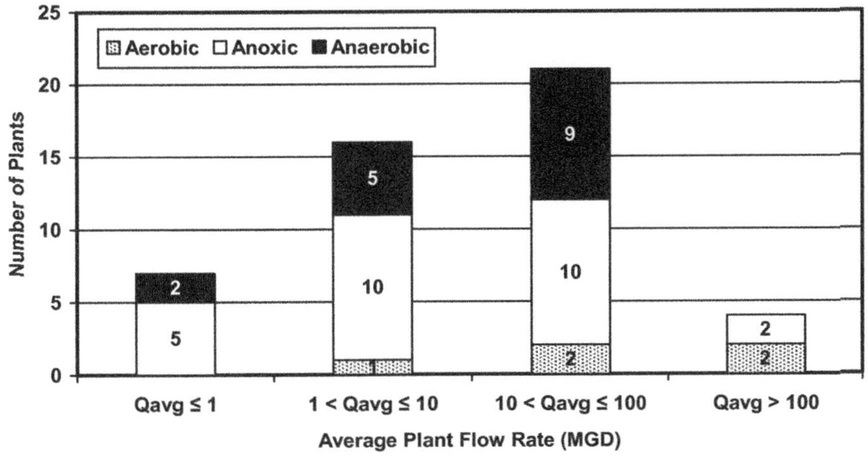

Figure 4-5. Facility Size, Selector Type Distribution.

4.4.2 Plant Flow vs. Settleability

Average plant flow is plotted against 90th percentile SVI and DSVI in Figure 4-6.

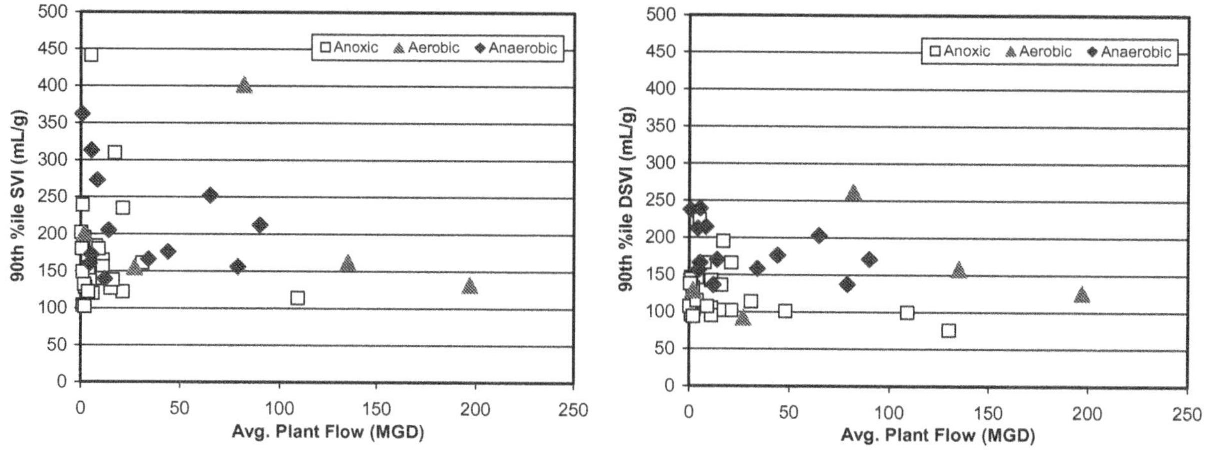

Figure 4-6. Plant Flow vs. 90th Percentile SVI and DSVI.

Table 4-5. Summary of Detailed Plant Investigation Data.

Plant Name	Location	Study Period	Avg. Flow (MGD)	Peak Flow (MGD)	Temp. Range (°C)	Reactor MCRT (d)	Selector MCRT (d)	Aerated MCRT (d)	No. of Selector Stages	Eff. No. of Selector Stages[3]
Aerobic										
King County South TP[1]	Renton, WA	Jan03-Aug04	82	154	13 - 23	2.8	0.3	2.8	2	2.6
Winston-Green WWTP[1]	Winston, OR	Jan03-Dec03	1.8	5.2	12 - 20	15	1.2	12	1	1.0
Upper Occoquan Sewage Authority	Centreville, VA	Jan04-Feb05	27	41	8 - 25	27	0.5	27	1	2.6
Deer Island WWTP Batteries A&B	Boston, MA	Apr04-Mar05	197	333	10 - 25	1.5	0.4	1.5	3	3.0
Deer Island WWTP Battery C	Boston, MA	Apr04-Mar05	135	175	10 - 25	1.4	0.3	1.4	2	2.3
Anoxic										
Springfield Northwest WWTP	Springfield, MO	Jan03-Dec03	3.8	7.4	9 - 25	12	2.1	9.7	1	1.1
Ashland WWTP	Ashland, OR	Nov03-Oct04	2.2	4.7	13 - 22	17	3.4	14	1	1.3
Bend WWTP	Bend, OR	Jan04-Dec04	5.1	6.4	14 - 24	10	3.3	7.0	3	3.0
Pierce County Chambers Creek WWTP	Univ. Place, WA	Jan03-May03	17	23	14 - 18	4.5	1.5	3.0	4	4.0
Olympus Terrace Sewer District WWTP	Mukilteo, WA	Jan04-Dec04	1.8	3.1	12 - 22	24	0.7	24	1	1.4
Winston-Green WWTP[1]	Winston, OR	Jan03-Dec03	0.9	1.6	15 - 25	16	1.2	14	1	1.0
Yakima WWTP	Yakima, WA	Jan03-Dec03	11	15	12 - 26	11	3.3	7.3	1	1.3
OMI - Plant 1.	Southeastern US	Jul03-Jul04	21	39	14 - 26	8.5	2.1	6.4	2	4.0
OMI - Plant 2	Southeastern US	Jul03-Jul04	0.82	2.4	11 - 25	38	7.7	26	3	3.0
OMI - Plant 3	Southwestern US	Jan03-Jul04	1.7	2.1	17 - 26	13	3.1	10	3	3.0
OMI - Plant 5	Pacific Northwest	Oct03-Oct04	0.88	3.8	11 - 21	11	3.0	8.1	3	3.5
OMI - Plant 6	South Central US	Jan00-Dec00	15	29	8.4 - 29	12	2.3	9.5	6	6.0
Veolia - Plant 1	Midwestern US	Jan04-Dec04	0.17	0.37	7 - 23	51	26	26	1	2.9
Veolia - Plant 4	New England	Mar04-Mar05	11	17	14 - 29	12	2.4	10	2	3.0
Veolia - Plant 11	New England	Jan04-Dec04	11	21	10 - 36	21	5.2	16	1	1.2
Veolia - Plant 12	South Central US	Jan04-Dec04	0.4	1.3	18 - 32	21	2.5	19	6	6.0
Landis Sewerage Authority	Vineland, NJ	--	5.9	--	--	15.2	--	--	3	--
Phoenix 91st Avenue WWTP	Tolleson, AZ	Jul03-Jun04	130	164	20 - 32	8.2	1.3	5.7	3	3.2
Puyallup WPCP	Puyallup, WA	May04-Apr05	3.7	10	12 - 23	24	9.1	10	4	4.0
City of Columbus Southerly WWTP	Lockbourne, OH	Jul03-Jun04	109	221	13 - 25	11	2.7	8.0	4	4.0
Davenport WPCP	Davenport, IA	Jun03-Nov04	21	41	10 - 25	21	2.0	19	3	3.0
Southside WWTP	Tulsa, OK	Jan04-Dec04	31	60	13 - 27	4.1	0.1	3.9	2	2.0
Northside WWTP[2]	Tulsa, OK	Jan04-Dec04	16	38	9 - 26	12	0.7	11	2	3.8
Phoenix 23rd Avenue WWTP	Phoenix, AZ	Jul03-Aug04	48	60	23 - 34	8.1	1.4	6.6	3	3.3
City of Brewer WPCF	Brewer, ME	Jan04-Dec04	1.9	5.1	5 - 21	11	1.1	9.9	1	1.0
Gilbert Neely WWRF	Gilbert, AZ	Oct03-Sep04	7.6	9.9	24 - 32	13	2.6	10	5	5.9
Glendale WWTF	Lakeland, FL	May04-Apr05	8.9	29	19 - 31	5.3	1.2	4.1	3	3.0
Anaerobic										
Snoqualmie WWTP	Snoqualmie, WA	--	0.61	2.4	8 - 22	16	2.4	14	3	--
King County South TP[1]	Renton, WA	Jan03-Aug04	90	185	12 - 24	3.1	0.3	2.8	2	2.6
OMI - Plant 4	Midwestern US	Jan02-Dec02	4.5	7.9	10 - 23	4.8	1.4	3.4	4	6.5
Veolia - Plant 8	Southeastern US	--	0.85	--	--	13	--	--	3	--
Tryon Creek WWTP	Portland, OR	Jan03-Aug04	8.2	21	12 - 21	3.8	0.8	3.0	2	2.4
Lynnwood WWTP	Lynnwood, WA	Jan04-Dec04	4.2	10.9	12 - 23	2.7	0.3	2.4	2	2.5
Columbia Blvd WWTP	Portland, OR	Sep03-Aug04	65	126	12 - 21	3.0	0.8	2.1	1	2.0
Southeast WPCP	San Francisco, CA	Feb03-Dec03	79	148	18 - 26	1.4	0.4	1.0	2	2.0
South Essex Sewerage District	Salem, MA	--	28	90	13 - 26	2.6	0.4	2.2	1	--
Central Contra Costa Sanitary District	Martinez, CA	Sep03-Aug04	44	82	18 - 27	1.5	0.2	1.3	1	5.9
Oro Loma Sanitary District	San Lorenzo, CA	Jan03-Dec03	14	28	15 - 24	3.3	0.5	2.8	1	1.1
Dublin San Ramon Sanitary District	Pleasanton, CA	Jan04-Dec04	12	20	20 - 28	2.3	0.4	1.9	1	2.4/3.0[4]
Billings WWTP (Train 2)	Billings, MT	Jan04-Dec04	5.3	8	11 - 20	3.6	1.1	2.6	2	3.4
Billings WWTP (Train 3)	Billings, MT	Jan04-Dec04	5.3	8	11 - 20	3.0	0.2	2.8	1	1.0
East Bay Municipal Utility District[2]	Oakland, CA	Jun03-Oct03	34	39	21 - 28	1.3	0.3	1.0	1	1.2
Orange County Sanitation District[2]	Fountain Valley, CA	Jul04-Nov04	29	36	27 - 30	1.6	0.3	1.4	1	1.2

Notes:

[1] Operation in this mode accounts for 50% of study period.

[2] Data repesents trains with selectors only.

[3] Calculated based on N number derived as part of this study (see Section 4.3.2.3).

[4] Values correspond to half- and full-compartment anaerobic selector operating modes, respectively.

Table 4-5. Summary of Detailed Plant Investigation Data (cont'd).

Plant Name	No. of Main Aeration Stages	Selector ICZ DO (mg/L)	Aeration Basin DO (mg/L)	Sec. Inf. Avg. BOD (mg/L)	MLSS (mg/L)	Contact Loading (mg BOD/g TSS)	Selector Loading F/ΣM [kg BOD5/(kg MLSS-d)]			
							ICZ	Stg. 2	Total	Eff. ICZ
Aerobic										
King County South TP[1]	8	--	2.0 - 2.5	156	1,800	59	7.2	4.2	--	10.9
Winston-Green WWTP[1]	2	--	3.6 - 6.8	141	3,800	48	1.8	--	--	1.8
Upper Occoquan Sewage Authority	2 - 3	0.5 - 1.5	1.3 - 3.3	167[3]	5,900	16[3]	2.5[3]	--	--	6.3 [3]
Deer Island WWTP Batteries A&B	4	--	8 - 20	83	1,600	49	15	7.5	3.8	15
Deer Island WWTP Battery C	4	--	9 - 20	83	1,500	52	8.0	4.0	--	9.3
Anoxic										
Springfield Northwest WWTP	2	0.5	0.6 - 5.9	230	2,300	38	0.39	--	--	0.44
Ashland WWTP	1	0.4	1.6 - 2.7	205[4]	3,500	--	0.18[4]	--	--	0.24 [4]
Bend WWTP	2	--	2.9 - 3.2	208	2,200	--	2.2	1.1	0.55	2.2
Pierce County Chambers Creek WWTP	2 - 3	--	1.8 - 3.0	137	2,100	66	6.6	3.5	0.87	6.6
Olympus Terrace Sewer District WWTP	2	--	0.2 - 2.9	184	3,300	54	2.6	--	--	3.6
Winston-Green WWTP[1]	4	< 0.5	1.6 - 5.6	208	3,900	68	1.4	--	--	1.4
Yakima WWTP	2 - 3	0.2	1.6	72	2,200	20	0.36	--	--	0.48
OMI - Plant 1	2	0.2	4.0 - 7.3	186	3,500	43	1.0	0.50	--	2.0
OMI - Plant 2	3	--	--	130	3,000	27	0.58	0.29	0.14	0.58
OMI - Plant 3	1	0.0	1.0 - 4.0	369	2,900	--	1.3	0.65	0.43	1.3
OMI - Plant 5	1	0.3	2 - 3	227	2,700	--	1.3	0.64	0.43	1.5
OMI - Plant 6	4	0	1.4 - 5.6	88	2,900	21	4.4	2.2	0.74	4.4
Veolia - Plant 1	1	0.2	0.1 - 9.7	407	5,100	80	0.095	--	--	0.22
Veolia - Plant 4	1	--	2.1 - 4.1	183	3,700	28	1.3	0.63	--	1.9
Veolia - Plant 11	2	0.1 - 0.3	2.0 - 3.8	246	4,100	28	0.59	--	--	0.68
Veolia - Plant 12	4	--	1.3 - 6.5	818	3,000	--	4.3	2.1	0.53	4.3
Landis Sewerage Authority	4	--	--	319[5]	2,400	--	1.1[5]	0.55[5]	0.37[5]	--
Phoenix 91st Avenue WWTP	8	--	1.1 - 2.0	149[6]	3,200	--	3.9[6]	2.0[6]	1.0[6]	3.9[6]
Puyallup WPCP	4	0.2	0.3 - 2.1	170[7]	3,000	--	2.2[7]	1.1[7]	0.16[7]	2.2[7]
City of Columbus Southerly WWTP	6	--	2.7 - 7.8	103	3,400	23	3.4	1.7	0.38	3.4
Davenport WPCP	3	0.7	0.5 - 1.0	169[8]	1500[9]	83[8]	4.9[8]	2.5[8]	1.6[8]	4.9[8]
Southside WWTP	1	0.5	2.2 - 5.1	259	3,100	43	21	10	--	21
Northside WWTP[2]	1	0.5	2.9 - 5.2	129	2,100	42	5.5	2.8	--	11
Phoenix 23rd Avenue WWTP	5	--	2.0 - 2.7	160	3,100	--	6.6	2.5	1.0	6.6
City of Brewer WPCF	1	0.2 - 1.2	3.2 - 8.6	105	3,600	21	1.0	--	--	1.1
Gilbert Neely WWRF	1	0.2	0.7 - 2.4	178	2,000	14	4.8	2.4	0.4	4.8
Glendale WWTF	4	0.2 - 0.8	0.7 - 2.8	413	4,300	--	4.4	2.2	1.2	4.4
Anaerobic										
Snoqualmie WWTP	1	--	0.4 - 1.3	250	1,600	--	1.2	0.59	0.23	--
King County South TP[1]	8	--	1.7 - 2.5	147	2,200	47	6.1	3.5	--	9.1
OMI - Plant 4	3	--	1.9 - 3.2	184	2,000	--	3.6	1.7	0.80	9.3
Veolia - Plant 8	1	0.5	0.7	238[10]	3,400	--	1.2[10]	0.6[10]	0.2[10]	--
Tryon Creek WWTP	3 - 5	--	1.5 - 2	82	1,500	--	2.4	1.2	--	2.9
Lynnwood WWTP	2	--	2.9	125	1,900	46	5.6	2.8	--	7.1
Columbia Blvd WWTP	1	0.1 - 0.5	0.1 - 4.8	243	2,000	90	2.1	--	--	4.1
Southeast WPCP	4	--	5.0 - 7.5	168	2,400	63	6.0	2.7	--	6
South Essex Sewerage District	4	--	--	100	1,500	58	3.0	--	--	--
Central Contra Costa Sanitary District	2	--	0 - 5	140	1,300	--	4.5	--	--	45
Oro Loma Sanitary District	6	--	--	123[11]	1,300	53[11]	2.0[11]	--	--	2.0
Dublin San Ramon Sanitary District	2 - 3	--	1.3 - 1.9	128	1,900	42	2.0	--	--	5.0
Billings WWTP (Train 2)	2	--	1.0 - 1.9	102[8]	2,000	41[8]	4.4[8]	1.1[8]	--	4.6[8]
Billings WWTP (Train 3)	2	--	1.0 - 1.5	102[8]	1,900	35[8]	4.5[8]	--	--	4.7[8]
East Bay Municipal Utility District[2]	3	--	--	230[12]	2,000	106[12]	5.1[12]	--	--	5.9[12]
Orange County Sanitation District[2]	5	--	1.0 - 2.5	172	800	149	5.7	--	--	6.5

Notes:

[1] Operation in this mode accounts for 50% of study period.
[2] Data repesents trains with selectors only.
[3] Reported on a COD basis. BOD_5 estimated as 0.5xCOD.
[4] Reported on a $cBOD_5$ basis. BOD_5 estimated as 1.1x$cBOD_5$ based on available plant data.
[5] Reported as plant influent BOD_5. Secondary influent BOD_5 estimated as 0.7 x plant influent BOD_5.
[6] Reported on a COD basis. BOD_5 estimated as 0.46xCOD based on available plant data.
[7] Reported on a COD basis. BOD_5 estimated as 0.45xCOD based on available plant data.
[8] Reported on a $cBOD_5$ basis. BOD_5 estimated to be equivalent to $cBOD_5$ based on available plant information.
[9] Reported value is from contact zone of contact-stabilization plant. Stabilitization zone MLSS is 6,000 mg/L.
[10] Reported on a $cBOD_5$ basis. No conversion to BOD_5 was made.
[11] Reported on a COD basis. BOD_5 estimated as 0.4xCOD based on available plant data.
[12] Reported on a $cBOD_5$ basis. BOD_5 estimated as 1.45 x $cBOD_5$ based on available plant data.

Table 4-5. Summary of Detailed Plant Investigation Data (cont'd).

	Avg. Selector HRT (w/o recycle) (h)					Avg. Selector HRT (w/recycle) (h)			Clarifier Underflow	Clarifier Underflow
Plant Name	**ICZ**	**Stg. 2**	**Stg.3-5**	**Total**	**Eff. ICZ**	**ICZ**	**Total**	**Eff. ICZ**	**Rate (%)**	**Range (%)**
Aerobic										
King County South TP[1]	0.3	0.2	--	0.5	0.2	0.21	0.36	0.14	40%	25-50%
Winston-Green WWTP[1]	0.6	--	--	0.6	0.6	0.36	0.36	0.36	55%	30-80%
Upper Occoquan Sewage Authority	0.2	--	--	0.2	0.1	0.14	0.14	0.05	75%	40-140%
Deer Island WWTP Batteries A&B	0.1	0.1	0.2	0.4	0.1	0.07	0.29	0.07	30%	25-35%
Deer Island WWTP Battery C	0.2	0.2	--	0.4	0.2	0.14	0.29	0.12	30%	30-50%
Anoxic										
Springfield Northwest WWTP	6.5	--	--	6.5	5.8	2.38	2.38	2.12	175%	125-200%
Ashland WWTP	7.8	--	--	7.8	5.9	3.99	3.99	3.02	--	35-40%
Bend WWTP	1.1	1.1	2.2	4.4	1.1	0.23	0.93	0.23	50%	--
Pierce County Chambers Creek WWTP	0.2	0.2	1.3	1.8	0.2	0.10	0.60	0.10	29%	20-35%
Olympus Terrace Sewer District WWTP	0.6	--	--	0.6	0.4	0.40	0.40	0.28	45%	20-80%
Winston-Green WWTP[1]	1.0	--	--	1.0	1.0	0.61	0.61	0.61	55%	40-70%
Yakima WWTP	2.4	--	--	2.4	1.8	1.42	1.42	1.06	70%	65-75%
OMI - Plant 1	1.3	1.3	--	2.6	0.7	0.94	1.89	0.48	40%	30-50%
OMI - Plant 2	1.9	1.9	4.0	7.8	1.9	1.17	4.80	1.17	65%	40-90%
OMI - Plant 3	2.3	2.3	2.3	6.9	2.3	0.27	0.82	0.27	35%	30-45%
OMI - Plant 5	1.7	1.7	1.7	5.1	1.5	0.47	1.423	0.41	30%	20-50%
OMI - Plant 6	0.2	0.2	0.7	1.0	0.2	0.13	0.81	0.13	25%	10-55%
Veolia - Plant 1	22	--	--	22	7.8	10.11	10.11	3.55	115%	--
Veolia - Plant 4	1.0	1.0	--	2.0	0.7	0.52	1.04	0.35	85%	60-110%
Veolia - Plant 11	2.6	--	--	2.6	2.3	0.52	0.52	0.45	110%	80-160%
Veolia - Plant 12	1.9	1.9	11	15	1.9	0.71	5.71	0.71	140%	--
Landis Sewerage Authority	4.1	4.1	4.1	12.3	--	--	--	--	--	--
Phoenix 91st Avenue WWTP	0.3	0.3	0.6	1.1	0.3	0.07	0.26	0.07	50%	50-55%
Puyallup WPCP	0.7	0.7	8.1	9.5	0.7	0.34	2.31	0.34	15%	10-25%
City of Columbus Southerly WWTP	0.3	0.3	1.8	2.4	0.3	0.18	1.58	0.18	50%	N/A
Davenport WPCP	0.6	0.6	0.6	1.9	0.6	0.47	1.40	0.47	35%	20-50%
Southside WWTP	0.1	0.1	--	0.2	0.1	0.06	0.11	0.06	75%	55-105%
Northside WWTP[2]	0.3	0.3	--	0.6	0.1	0.18	0.36	0.09	60%	40-80%
Phoenix 23rd Avenue WWTP	0.2	0.3	0.7	1.2	0.2	0.06	0.31	0.06	50%	45-50%
City of Brewer WPCF	0.7	--	--	0.7	0.7	0.46	0.46	0.44	55%	35-80%
Gilbert Neely WWRF	0.4	0.4	4.4	5.3	0.4	0.07	0.86	0.07	160%	110-200%
Glendale WWTF	0.6	0.6	0.9	2.0	0.6	0.16	0.43	0.16	60%	30-120%
Anaerobic										
Snoqualmie WWTP	3.2	3.2	10	16.4	--	--	--	--	--	--
King County South TP[1]	0.3	0.2	--	0.5	0.2	0.20	0.34	0.13	39%	30-55%
OMI - Plant 4	0.6	0.7	1.5	2.8	0.2	0.39	0.86	0.15	65%	40-80%
Veolia - Plant 8	1.4	1.4	5.7	8.5	--	--	--	--	--	--
Tryon Creek WWTP	0.6	0.6	--	1.2	0.5	0.47	0.94	0.39	25%	10-45%
Lynnwood WWTP	0.3	0.3	--	0.6	0.2	0.23	0.45	0.18	35%	25-50%
Columbia Blvd WWTP	1.5	--	--	1.5	0.7	1.11	1.11	0.56	35%	25-45%
Southeast WPCP	0.3	0.4	--	0.7	0.3	0.24	0.53	0.24	30%	25-35%
South Essex Sewerage District	0.5	--	--	0.5	--	--	--	--	45%	--
Central Contra Costa Sanitary District	0.6	--	--	0.6	0.1	0.42	0.42	0.04	40%	30-45%
Oro Loma Sanitary District	1.2	--	--	1.2	1.0	0.60	0.60	0.52	95%	70-110%
Dublin San Ramon Sanitary District	1.0	--	--	1.0	0.4	0.58	0.58	0.22	65%	50-80%
Billings WWTP (Train 2)	0.3	0.9	--	1.2	0.3	0.21	0.85	0.20	45%	40-60%
Billings WWTP (Train 3)	0.3	--	--	0.3	0.3	0.19	0.19	0.18	60%	50-70%
East Bay Municipal Utility District[2]	0.6	--	--	0.6	0.5	0.43	0.43	0.37	30%	25-35%
Orange County Sanitation District[2]	1.0	--	--	1.0	0.8	0.55	0.55	0.48	75%	60-90%

Notes:

[1] Operation in this mode accounts for 50% of study period.

[2] Data repesents trains with selectors only.

Table 4-5. Summary of Detailed Plant Investigation Data (cont'd).

Plant Name	Avg. SVI (mL/g)	90th %ile SVI (mL/g)	% of SVIs > 150 (mL/g)	Avg. Merkel DSVI (mL/g)	90th %ile Merkel DSVI (mL/g)
Aerobic					
King County South TP[1]	196	402	50%	164	262
Winston-Green WWTP[1]	128	201	36%	97	129
Upper Occoquan Sewage Authority	96	155	12%	66	92
Deer Island WWTP Batteries A&B	100	132	5%	99	126
Deer Island WWTP Battery C	116	163	14%	112	158
Anoxic					
Springfield Northwest WWTP	129	154	13%	124	146
Ashland WWTP	89	128	4%	81	103
Bend WWTP	243	441	77%	168	224
Pierce County Chambers Creek WWTP	249	310	99%	178	195
Olympus Terrace Sewer District WWTP	112	132	2%	101	114
Winston-Green WWTP[1]	77	104	2%	74	96
Yakima WWTP	121	165	15%	114	142
OMI - Plant 1	98	122	2%	87	102
OMI - Plant 2	187	239	80%	128	146
OMI - Plant 3	127	195	29%	105	132
OMI - Plant 5	108	149	9%	107	144
OMI - Plant 6	91	127	1%	83	102
Veolia - Plant 1	174	202	91%	92	107
Veolia - Plant 4	119	156	12%	94	105
Veolia - Plant 11	93	140	8%	76	95
Veolia - Plant 12	131	180	20%	110	138
Landis Sewerage Authority	92	120	0%	--	--
Phoenix 91st Avenue WWTP	N/A	N/A	N/A	71[3]	76[3]
Puyallup WPCP	106	122	0%	99	115
City of Columbus Southerly WWTP	100	114	0%	88	100
Davenport WPCP	128	235	18%	105	166
Southside WWTP	101	161	12%	90	114
Northside WWTP[2]	114	138	4%	113	136
Phoenix 23rd Avenue WWTP	N/A	N/A	N/A	81[3]	101[3]
City of Brewer WPCF	75	102	0%	74	95
Gilbert Neely WWRF	116	184	17%	112	166
Glendale WWTF	96	180	16%	77	107
Anaerobic					
Snoqualmie WWTP	273	362	94%	208	237
King County South TP[1]	147	212	26%	130	171
OMI - Plant 4	127	160	20%	126	156
Veolia - Plant 8	147	--	--	--	--
Tryon Creek WWTP	189	273	73%	171	214
Lynnwood WWTP	N/A	N/A	N/A	147[3]	212[3]
Columbia Blvd WWTP	170	252	62%	153	203
Southeast WPCP	120	156	13%	112	137
South Essex Sewerage District	108	--	--	--	--
Central Contra Costa Sanitary District	139	176	24%	139	176
Oro Loma Sanitary District	147	205	33%	132	170
Dublin San Ramon Sanitary District	100	139	7%	98	136
Billings WWTP (Train 2)	153	172	48%	144	166
Billings WWTP (Train 3)	222	313	84%	182	239
East Bay Municipal Utility District[2]	120	166	19%	117	158
Orange County Sanitation District[2]	518	731	100%	420	579

Notes:

[1] Operation in this mode accounts for 50% of study period.

[2] Data repesents trains with selectors only.

[3] Data reported by plants on a DSVI basis. No Merkel equation correction was applied.

Table 4-5. Summary of Detailed Plant Investigation Data (cont'd).

Plant Name	Dominant Filamentous Organisms	Other Observed Filamentous Organisms
Aerobic		
King County South TP[1]	1701, 021N	
Winston-Green WWTP[1]	*N. limicola*	
Upper Occoquan Sewage Authority	*M. parvicella*	*N. limicola, 0041, 1851*
Deer Island WWTP Batteries A&B	*Thiothrix* , 021N, *S. natans* [3]	
Deer Island WWTP Battery C	*Thiothrix, 021N, S. natans* [3]	
Anoxic		
Springfield Northwest WWTP	*N. limicola, S. natans, M. parvicella* [3]	
Ashland WWTP	0675, 0041	
Bend WWTP	*M. parvicella*	*S. natans*
Pierce County Chambers Creek WWTP	1863	*Thiothrix, S. natans,* 1701
Olympus Terrace Sewer District WWTP	0675, 0041	*N. limicola II,* 0803, 1851
Winston-Green WWTP[1]	*N. limicola*	
Yakima WWTP	*S. natans* , 021N, *Thiothrix I/II*	
OMI - Plant 1	0675	*H. hydrossis,* 0041
OMI - Plant 2	0675, *H. hydrossis*	0041, *Thiothrix II*
OMI - Plant 3	1851, *M. parvicella*	*N. limicola I* , 0961, *H. hydrossis,* 0092, 0914
OMI - Plant 5	1851	0041, *N. limicola II, H. hydrossis*
OMI - Plant 6	021N, *Thiothrix*	
Veolia - Plant 1	*M. parvicella* , 0041, 0675, 1851	
Veolia - Plant 4	No dominant filaments identified.	0041, 0675, *N. limicola II,* 1701
Veolia - Plant 11	0041, 0675	1701, *N. limicola II, M. parvicella,* 0092
Veolia - Plant 12	0675, *N. limicola I,* 0041	*H. hydrossis*
Landis Sewerage Authority	0041	
Phoenix 91st Avenue WWTP	*Thiothrix II* , 0092, 0675, 1701, 0041	*0914, Thiothrix I*
Puyallup WPCP	0041	*Thiothrix II,* 0675
City of Columbus Southerly WWTP	021N	
Davenport WPCP	021N, *N. limicola I/II*	0914, *Thiothrix I/II*
Southside WWTP	No dominant filaments identified.	0092, 0675, 0041, *H. hydrossis*
Northside WWTP[2]	0581	*Thiothrix I,* 0914, *M. parvicella,* 0041
Phoenix 23rd Avenue WWTP	1701, *Thiothrix II* , 1851, 0675, 0041, 0914	021N, 0803
City of Brewer WPCF	*N. limicola I*	*M. parvicella,* 0914
Gilbert Neely WWRF	0092, *M. parvicella* [3]	
Glendale WWTF	*N. limicola II/III*	
Anaerobic		
Snoqualmie WWTP	*M. parvicella*	
King County South TP[1]	1863	
OMI - Plant 4	*M. parvicella*	*N.limicola II,* 0041
Veolia - Plant 8	Not available	
Tryon Creek WWTP	1701, *N. limicola II*	*H. hydrossis*
Lynnwood WWTP	1863, 1701 [3]	
Columbia Blvd WWTP	1701	*H. hydrossis, N. limicola II,* 0675
Southeast WPCP	No dominant filaments identified.	0041, 0675, *N. limicola II*
South Essex Sewerage District	*S. natans*	
Central Contra Costa Sanitary District	021N	
Oro Loma Sanitary District	No dominant filaments identified.	1701, *H. hydrossis*
Dublin San Ramon Sanitary District	*Thiothrix* , 0914, 0041, 0675, *H. hydrossis* , *N. limicola* , 1863 [3]	
Billings WWTP (Train 2)	*M. parvicella*	*N. limicola II,* 021N
Billings WWTP (Train 3)	*M. parvicella,* 1701	*N. limicola II, H. hydrossis*
East Bay Municipal Utility District[2]	021N, *S. natans* , 1701	*H. hydrossis, N. limicola*
Orange County Sanitation District[2]	021N, *Thiothrix, S. natans,* 1701	

Notes:
[1] Operation in this mode accounts for 50% of study period.
[2] Data repesents trains with selectors only.
[3] Multiple filaments reported, but relative dominance unknown.

4.4.3 Selector ICZ F/M vs. Settleability

As discussed in Section 2.7, the literature identifies the F/M loading applied to the selector ICZ as a key parameter in selector design and performance. Suggested design ICZ F/M loading rates were presented in Table 2-4. Figure 4-7 is a plot of selector ICZ F/M versus 90th percentile SVI and DSVI. A wide range of settling performance was achieved across a broad range of ICZ F/M loading rates. Most of the anaerobic selector DSVI values exceeded the typical control limit of 150 mL/g; conversely, the majority of anoxic selector installations exhibited acceptable bulking control. No trends were observed between selector ICZ F/M loading and settling performance within any of the selector classifications.

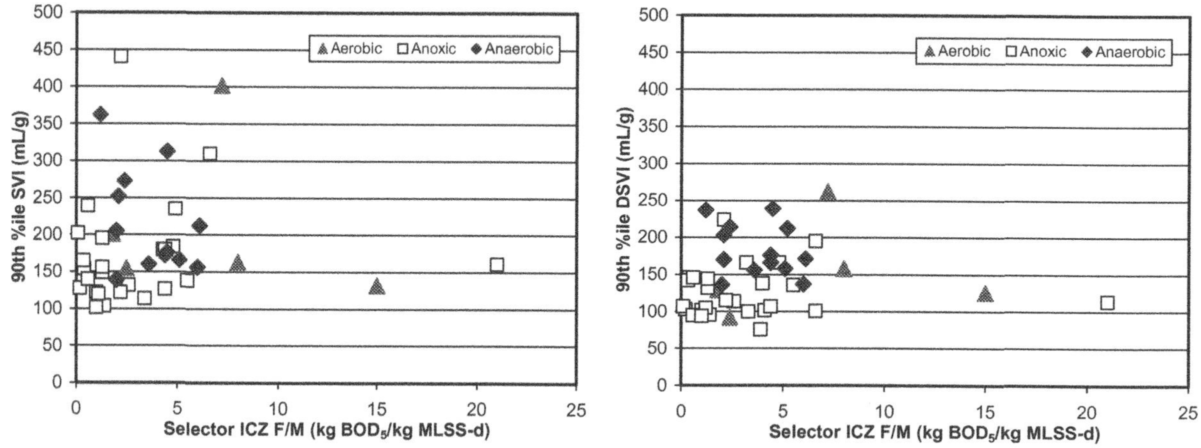

Figure 4-7. Selector ICZ F/M vs. 90th Percentile SVI and DSVI.

As discussed in Section 4.3.2.3, the effective number of compartments in the selector zone was estimated based on a semi-empirical formula. The revised number of selector compartments was used to recalculate the ICZ volume and the associated selector ICZ F/M loading. No trends were observed in the plot of effective selector ICZ F/M versus 90th percentile SVI and DSVI (Figure 4-8).

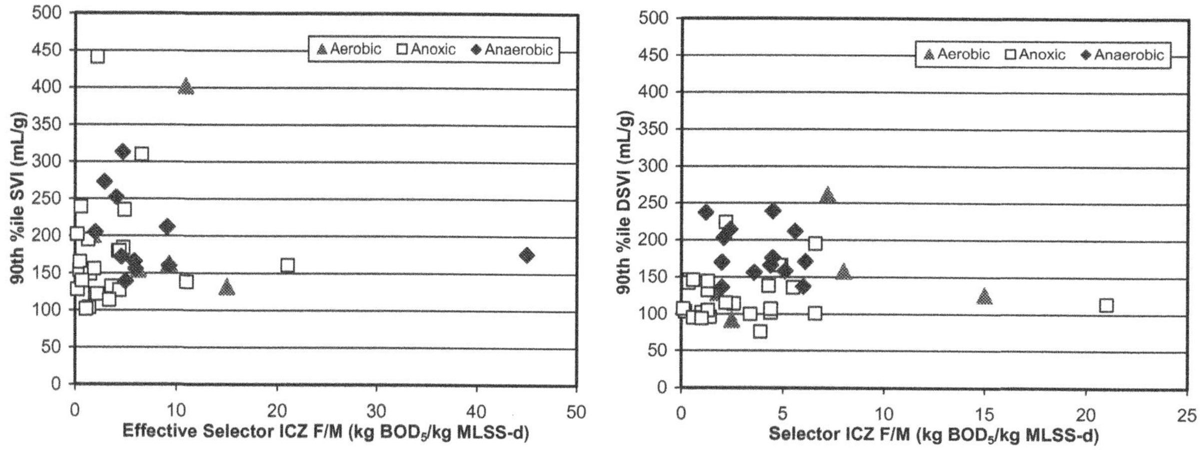

Figure 4-8. Effective Selector ICZ F/M vs. 90th Percentile SVI and DSVI.

4.4.4 Total Selector F/M vs. Settleability

Figure 4-9 is a comparison of total selector F/M (mixed liquor in total selector zone only) and 90th percentile SVI and DSVI. As discussed in Section 2.7.3, the literature suggests total selector F/M loadings of 1.5–3 kg COD/(kg MLSS·d) for anoxic and anaerobic selectors and 3 kg COD/(kg MLSS·d) for aerobic selectors. Figure 4-7 clearly illustrates that the majority of anoxic selectors are operated below the recommended F/M range, yet they are able to control DSVIs in most cases. The aerobic and anaerobic selectors were operated over a wide range of F/M values with significant variations in settleability.

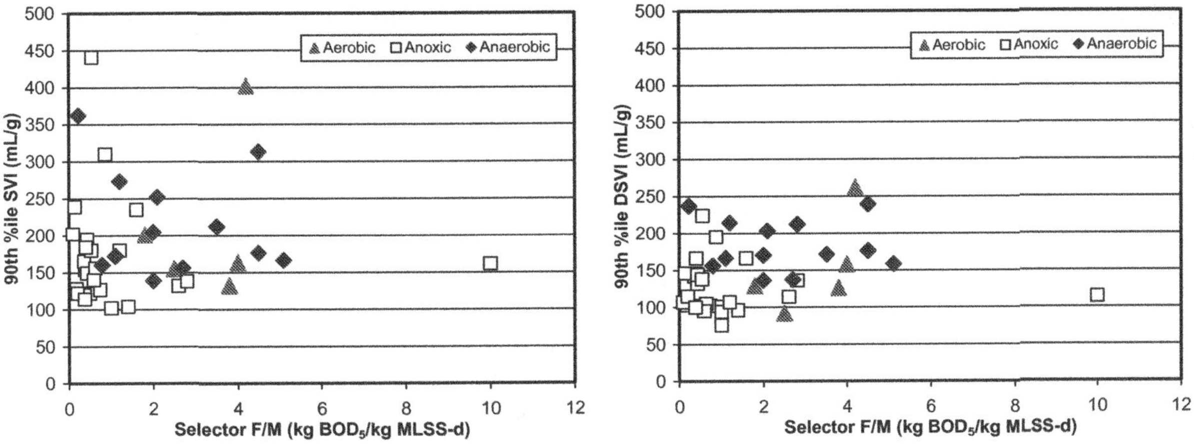

Figure 4-9. Total Selector F/M vs. 90th Percentile SVI and DSVI.

4.4.5 Selector MCRT vs. Settleability

Figure 4-10 is a comparison of selector MCRT (mixed liquor in selector zone only) and settleability. Nearly all anaerobic selector installations are operated at a selector MCRT less than 2.0 d, while values for anoxic selectors are distributed across a range of 0.3–9.1 d. At a selector MCRT greater than 2.0 d, only one of 12 anoxic selectors produced DSVIs greater than 150 mL/g, suggesting improved bulking control at higher selector MCRTs. Anaerobic selectors performed poorly across a range of selector MCRTs with only two of nine facilities demonstrating acceptable DSVI control. This analysis is based on the 90th percentile SVI and DSVI and an average selector MCRT value to represent each plant. Therefore, this method may exclude important data that could affect the outcome of this analysis.

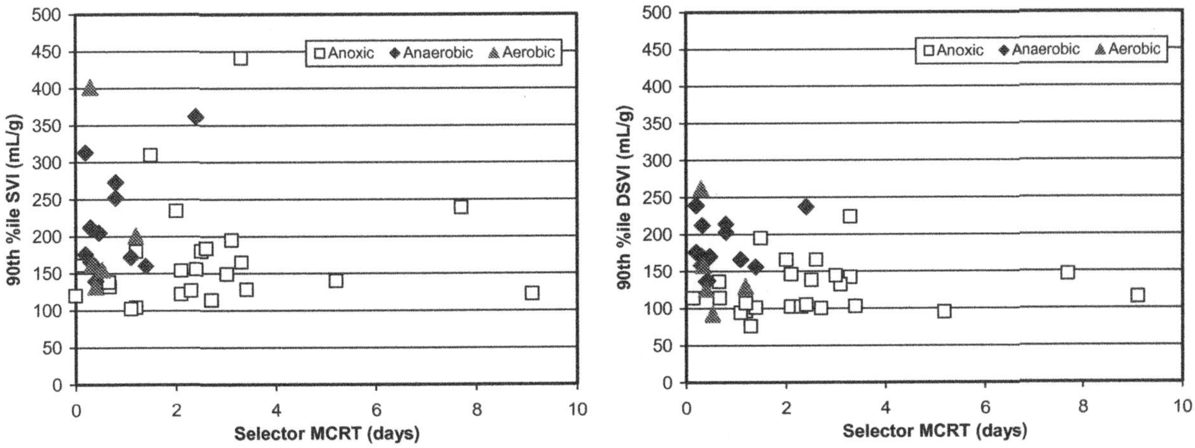

Figure 4-10. Selector MCRT vs. 90th Percentile SVI and DSVI.

4.4.6 Reactor MCRT vs. Settleability

System MCRT (excluding clarifier solids) versus 90th percentile SVI and DSVI is presented in Figure 4-11. The anoxic selectors achieved DSVI control over a wide range of system MCRTs with a cluster of plants performing poorly at a system MCRT of 2.7–4.2 d. Nearly all anaerobic selectors with a system MCRT less than 4.8 d performed poorly.

This figure also demonstrates how plants with an MCRT <5 d tend to have anaerobic selectors, while plants with longer MCRTs tend to have anoxic selectors. Higher nitrate/nitrite concentrations may be present in the mixed liquor and/or RAS recycle streams in long-MCRT plants because of the higher likelihood that these plants will be completely nitrifying. The question is whether anoxic selectors perform better than anaerobic selectors, or do long-MCRT plants tend to have lower DSVIs than short-MCRT plants. Lower DSVIs may be a characteristic of long- versus short-MCRT plants. Appendix D, "Percentile Distribution Analysis of Regression Analysis Data Sets," discusses this topic further.

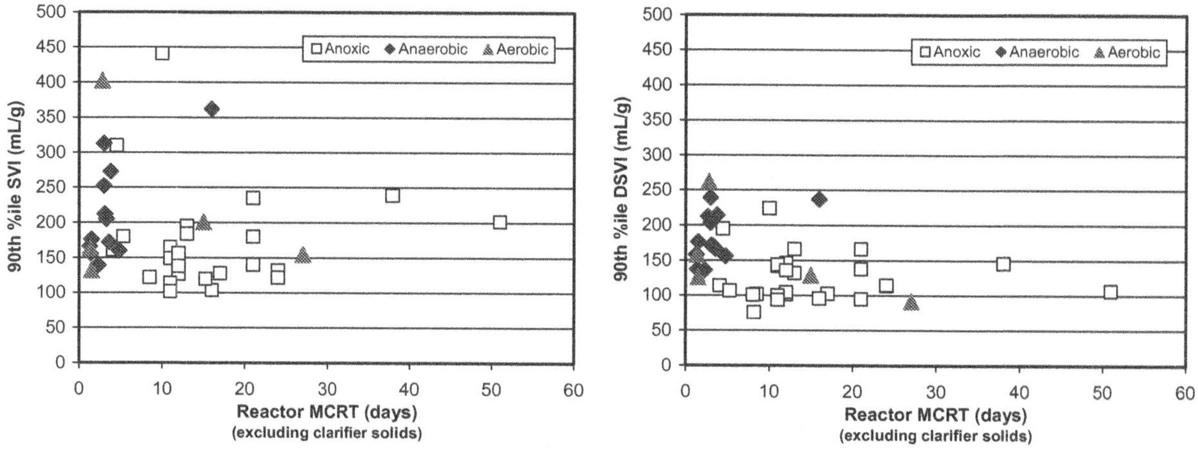

Figure 4-11. Reactor MCRT (excluding clarifier solids) vs. 90th Percentile SVI and DSVI.

4.4.7 Contact Loading vs. Settleability

Albertson (2005) recommended a contact loading maximum limit of 100 mg BOD_5/g MLSS be established to prevent overloading floc-formers in the selector zone. Figure 4-12 indicates that average contact loading conditions for nearly all study plants were well below the upper contact loading limit. Although two of the anoxic selector plants performed poorly at a contact loading greater than 80 mg BOD_5/g MLSS, no apparent trend was observed between contact loading and settling performance. Sufficient data to calculate contact loading was available for 33 of the 48 data sets.

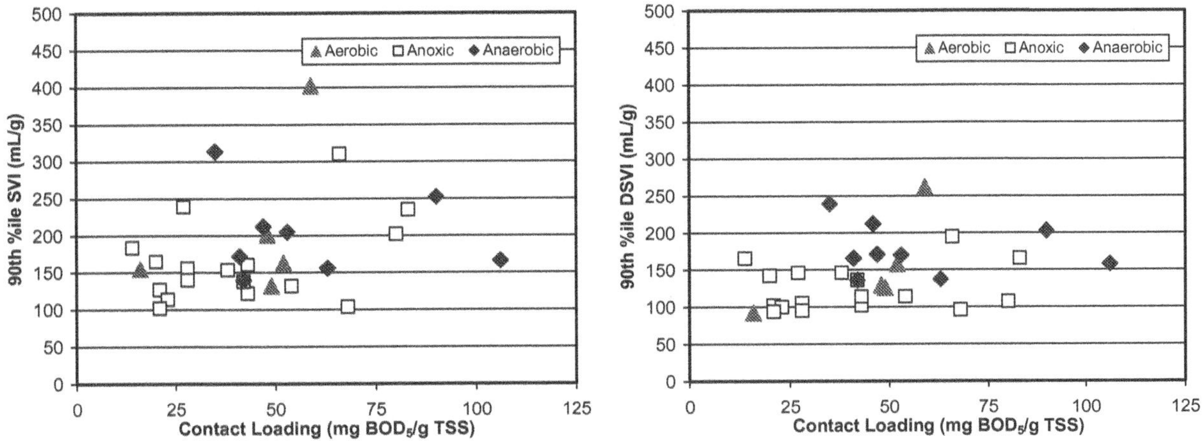

Figure 4-12. Contact Loading vs. 90th Percentile SVI and DSVI.

4.4.8 Total Selector HRT vs. Settleability

A typical recommended design guideline for the total selector HRT is approximately 0.75–2.0 h; however, Figure 4-13 clearly illustrates that many selector installations did not perform well even within this selector HRT range.

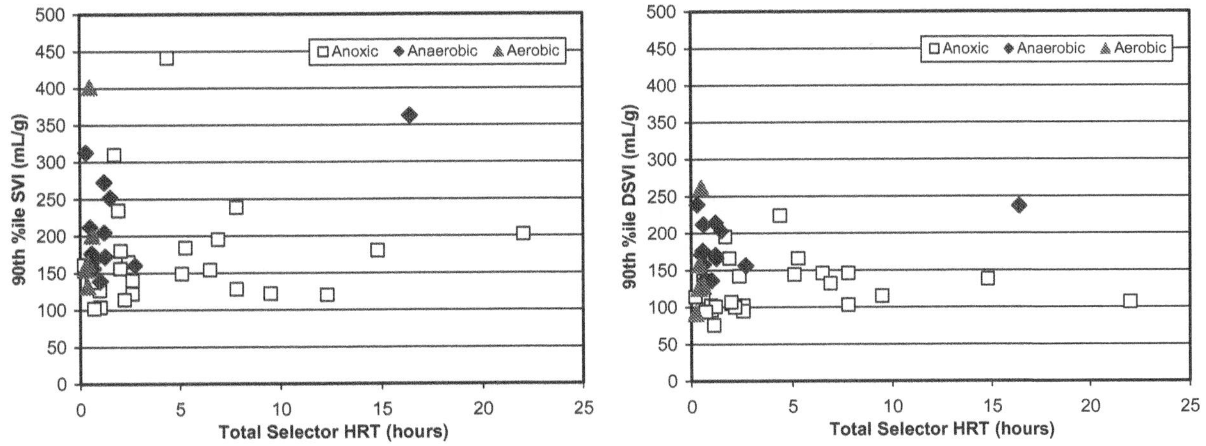

Figure 4-13. Total Selector HRT vs. 90th Percentile SVI and DSVI.

4.4.9 Ratio of Selector ICZ to Total Selector Volume vs. Settleability

The ratio of the selector ICZ volume to the total selector volume ($V_{S\text{-}ICZ}/V_{S\text{-}Tot}$) is a relative expression of selector staging and configuration, which may be related to F/M gradient loading conditions. For example, a three-stage selector with stage volumes of 25%, 25%, and 50% of total selector volume will have a $V_{S\text{-}ICZ}/V_{S\text{-}Tot}$ of 25% and cascade loading conditions of 4x, 2x, and x kg BOD_5/kg MLSS-d, respectively. Figure 4-14 indicates that selector configuration does not appear to be well correlated to SVI control. Similar ranges of SVI control were achieved for single-stage (represented by $V_{S\text{-}ICZ}/V_{S\text{-}Tot} = 100\%$) and multi-stage selectors with a wide variety of ICZ relative volumes. In fact, four single-stage anoxic selector installations exhibited acceptable DSVI control, while five multi-stage anoxic selectors with ICZ relative volumes ranging from 25%-50% did not perform well.

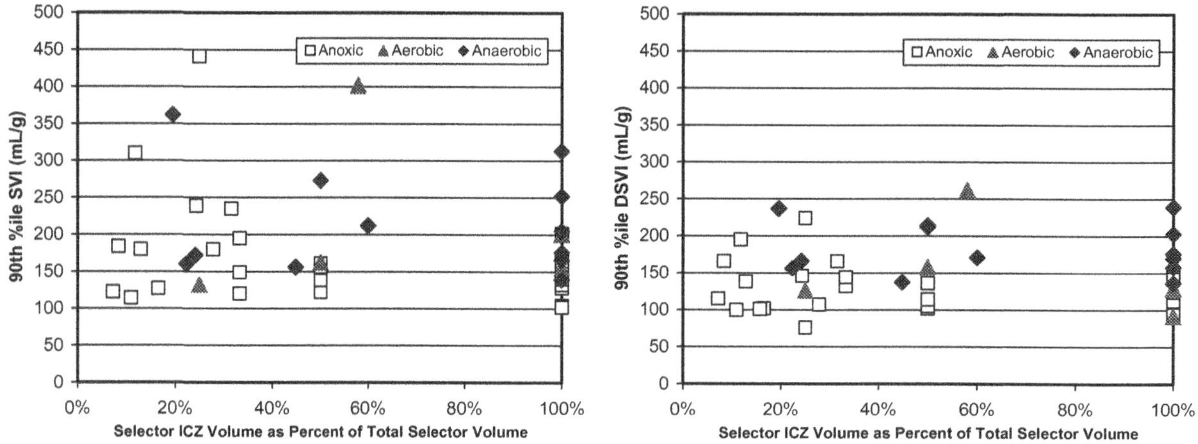

Figure 4-14. Ratio of Selector ICZ Volume to Total Selector Volume vs. 90th Percentile SVI and DSVI.

4.4.10 Number of Selector Stages vs. Settleability

Figure 4-15 is a comparison between SVI control achieved by single- vs. multi-staged selector installations. Three out of four plants equipped with single-stage anaerobic selectors had 90th percentile DSVIs greater than 150 mL/g, which is the same result for two-stage anaerobic selectors. Only four of 27 multi-staged anoxic selectors did not achieve acceptable DSVI control.

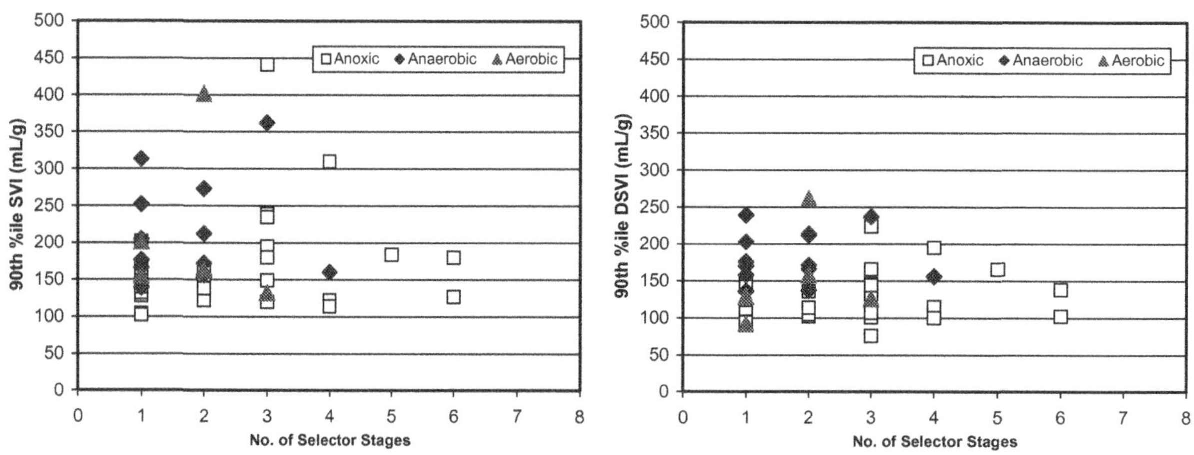

Figure 4-15. Number of Selector Stages vs. 90th Percentile SVI and DSVI.

Figure 4-16 compares the effective number of selector stages, calculated as discussed in Section 4.3.2.3, to settleability control.

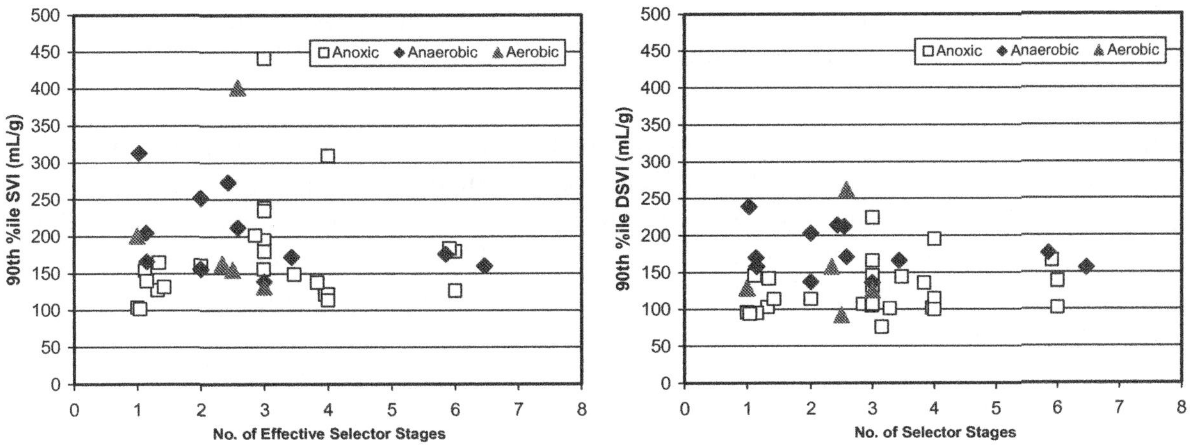

Figure 4-16. Number of Effective Selector Stages vs. 90th Percentile SVI and DSVI.

4.4.11 MLSS vs. Settleability

Figure 4-17 is a plot of MLSS versus 90th percentile SVI and DSVI. All facilties operating at an MLSS >2,500 mg/L yielded 90th percentile DSVIs <150 mL/g (primarily anoxic selector systems). Many of the facilities operating at an MLSS <2,500 mg/L yielded DSVIs >150 mL/g (mainly anaerobic selector plants).

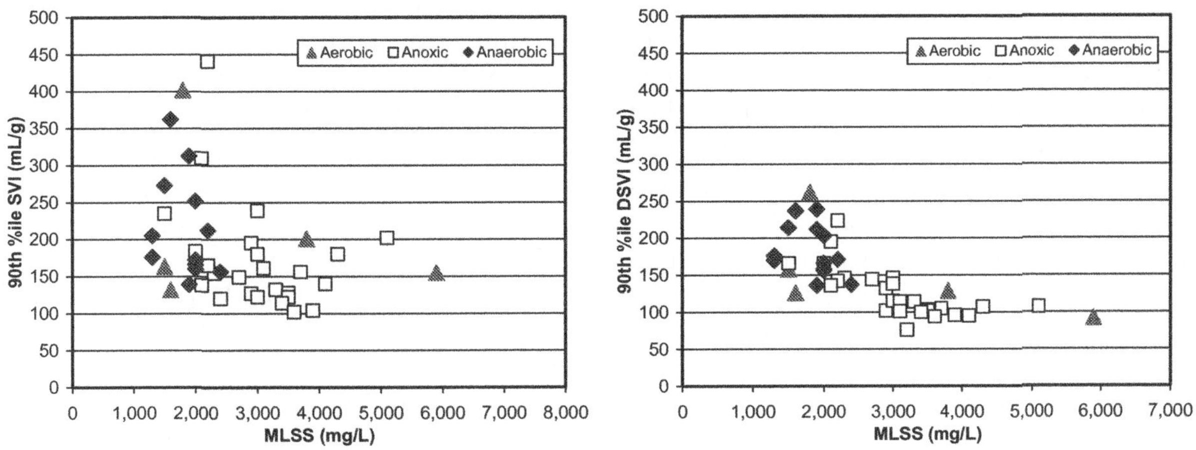

Figure 4-17. MLSS vs. 90th Percentile SVI and DSVI.

As discussed in Section 4.3.2.4, comparing average parameter and 90th percentile SVI/DSVI values for the plants included in the detailed plant investigation is somewhat limited because each facility is represented by only a single data point and does not consider how the variation in each parameter may influence variation in DSVI. Further, the plants were not separated by short and long MCRT, which may have provided different results. A single-variable regression analysis, incorporating daily operating data for each facility, was conducted to better evaluate the influence of parameter variation on SVI and DSVI values.

4.4.12 Regression Analysis

The purpose of the regression analysis was to evaluate design and operating parameters for activated sludge selectors that are frequently discussed in the literature and to quantify their relative influence on activated sludge settling characteristics as measured by sludge volume indices (either SVI or DSVI). The regression analysis was not intended to develop a regression model to calculate SVI.

Because selector design and operating parameters are often calculated using measured values that are also used to calculate other selector design and operating parameters—for example, activated sludge influent BOD_5 is used to calculate the influent BOD_5/TSS ratio, ICZ F/M, total selector F/M, and total system F/M, while the MLSS is used to calculate those F/Ms, selector MCRT, and total system MCRT, etc.—multicollinearity between the parameters was expected. Multicollinearity can distort regression analysis t-statistics used to measure relative parameter/independent variable influence on the dependent variable (DSVI). Some multicollinearity is tolerable.

The variance inflation factor (VIF) is used to measure multicollinearity. A VIF = 1.0 denotes no multicollinearity, and a VIF >10 denotes substantial multicollinearity and that severe distortion of the regression analysis results is likely (Tsai, 2001). A VIF <5.0 is usually considered tolerable with little significant impact to the regression results (Tsai, 2001; Montgomery and Peck, 1982). Extensive regression trials were performed to discover the best regression fit to the field data collected. Even though VIFs were kept below 5.0 for all parameters tested, it became apparent that multicollinearity between the parameters was so severe that multiple regression analysis of these parameters could not be valid.

Tables 4-6 to 4-8 show regression results for parameters tested using long-MCRT WWTP data. These results illustrate the severe effects of multicollinearity on the regression t-statistics (T). If the t-statistics listed in the tables are greater than 2 or less than -2, then there is a significant relationship between the independent variable ("predictor" in the tables) and the dependent variable (DSVI) (DeLurgio, 1998). If the t-statistic is negative, the dependent variable decreases when the independent variable increases, and if the t-statistic is positive, the dependent variable increases when the independent variable increases. If the p-value (P) listed in the tables is less than 1%, there is "overwhelming evidence" that that the regression relation between the dependent and independent variable is valid and highly significant (Keller and Warrack, 2000). Table 4-6 shows the regression results when four parameters were tested: 1) number of aeration basin stages, 2) average MLSS, 3) selector volume to total basin volume ratio, and 4) 7-d average selector MCRT >1 d (a dichotomous variable where 0.0 denotes MCRTs ≤1 d and 1.0 denotes MCRTs >1 d).

Table 4-6. Regression Analysis Trial (A) of Selector Parameters vs. Log DSVI: Long-MCRT WWTPs.

Predictor	T	P	VIF
Constant	487.04	0.000	
No. of Aeration Basin Stages	-4.86	0.000	1.1
Avg MLSS	-70.68	0.000	1.1
Selector Vol./Total Basin Vol.	29.78	0.000	1.5
7-d Avg. Selector MCRT >1 d	-15.77	0.000	1.5

R-Sq = 42.4%; 9,898 cases used; 766 cases contain missing values

Table 4-7 shows the same regression analysis but without the selector volume to total basin volume ratio parameter. Comparing Tables 4-6 and 4-7 shows the dramatic change in t-statistic values for the "number of aeration basin stages" parameter (from -4.85 to -15.17) and the "7-d average selector MCRT >1d" parameter (from -15.77, indicating strong influence on DSVI, to -0.22, indicating no significant influence on DSVI) when one parameter is removed from the regression analysis.

Table 4-7. Regression Analysis Trial (B) of Selector Parameters vs. Log DSVI: Long-MCRT WWTPs.

Predictor	T	P	VIF
Constant	570.58	0.000	
No. of Aeration Basin Stages	-15.17	0.000	1.0
Avg MLSS	-71.61	0.000	1.1
7-d Avg. Selector MCRT >1 d	-0.22	0.825	1.1

R-Sq = 37.2%; 9,898 cases used; 766 cases contain missing values

Table 4-8 shows an even more dramatic distortion of the "number of aeration basin stages" parameter t-statistic when it changes from significantly negative (-15.17) to significantly positive (8.71). The t-statistics in Table 4-6 indicate that the DSVI *decreases* with more aeration basin stages, but the t-statistics in Table 4-8 show that the DSVI *increases* with more aeration basin stages. Exactly the same data was used in both regression analyses (Tables 4-6 and 4-8). The t-statistic for the "7-d average selector MCRT >1 d" also increases sharply from -0.22 to -13.87, implying that this parameter is again strongly significant.

Table 4-8. Regression Analysis Trial (C) of Selector Parameters vs. Log DSVI: Long-MCRT WWTPs.

Predictor	T	P	VIF
Constant	283.98	0.000	
Temp (°C)	-25.17	0.000	1.4
No of Aeration Basin Stages	8.71	0.000	2.8
Selector Vol./Total Basin Vol.	23.81	0.000	2.0
7-d Avg. Selector MCRT >1 d	-13.87	0.000	1.9
7-d Avg. Reactor MCRT (d)	5.25	0.000	1.6
Sx1 Anoxic	-5.98	0.000	2.3
Sx1 Anaerobic	-1.71	0.088	3.4
Plant Avg. Flow (mgd)	-12.19	0.000	2.2
Avg MLSS (mg/L)	-62.18	0.000	1.7

R-Sq = 49.6%; 9,261 cases used; 1,403 cases contain missing values

It became apparent that a multiple regression analysis approach, even when restricting the analysis to only the most significant design and operating selector parameters, would not be valid. Since the goal of the regression analysis was to evaluate selector design and operating parameters based on their influence on DSVI and not to develop a model, each parameter was regressed against DSVI separately to provide truer t-statistics, unaffected by multicollinearity. Further, since R^2 is defined as the percent variation observed in DSVI (dependent variable) that is explained by the variation in the selector design/operating parameter (independent variable) (DeLurgio, 1998; Keller and Warrack, 2000), the R^2 value can be used to rank parameters according to their influence on DSVI.

Short-MCRT WWTP data was analyzed separately from long-MCRT WWTP data because it has been shown that filamentous bacteria dominating activated sludges in short-MCRT systems have significantly different growth requirements than those filamentous bacteria dominant in long-MCRT systems. Further, selectors are not as effective in controlling long-MCRT filaments (Gabb, 1988; Gabb et al., 1991; Wanner, 1994; Jenkins et al., 2004; Martins et al., 2004b). Because aerobic selectors primarily rely on kinetic mechanisms while anoxic and anaerobic selectors may use a combination of metabolic and kinetic mechanisms, short-MCRT plants with aerobic selectors were separated from short-MCRT plants with unaerated selectors in the analysis. Tables 4-9, 4-10, and 4-11 show how each of the facilities studied was separated into either short-MCRT plants with anoxic or anaerobic selectors, short-MCRT plants with aerobic selectors, or long-MCRT plants, respectively.

Table 4-9. Short-MCRT Plants with Anoxic or Anaerobic Selectors.

Plant Name	Location	Study Period	Avg. Flow (MGD)	Peak Flow (MGD)	Temp. Range (°C)	System MCRT (d)	Selector MCRT (d)	Aerated MCRT (d)	No. of Selector Stages	Eff. No. of Selector Stages	No. of Main Aeration Stages	Selector ICZ DO (mg/L)	Aeration Basin DO (mg/L)	Sec. Inf. Avg. BOD (mg/L)	MLSS (mg/L)	Contact Loading (mg BOD/g TSS)
Anoxic																
Pierce County Chambers Creek WWTP	Univ. Place, WA	Jan03-May03	17	23	14 - 18	4.5	1.5	3.0	4	4.0	2 - 3	--	1.8 - 3.0	137	2,100	66
Yakima WWTP	Yakima, WA	Jan03-Dec03	11	15	12 - 26	11	3.3	7.3	1	1.3	2 - 3	0.2	1.6	72	2,200	20
OMI - Plant 6	South Central US	Jan00-Dec00	15	29	8.4 - 29	12	2.3	9.5	6	6.0	4	0	1.4 - 5.6	88	2,900	21
Anaerobic																
King County South TP[1]	Renton, WA	Jan03-Aug04	90	185	12 - 24	3.1	0.3	2.8	2	2.6	8	--	1.7 - 2.5	147	2,200	47
OMI - Plant 4	Midwestern US	Jan02-Dec02	4.5	7.9	10 - 23	4.8	1.4	3.4	4	6.5	3	--	1.9 - 3.2	184	2,000	--
Tryon Creek WWTP	Portland, OR	Jan03-Aug04	8.2	21	12 - 21	3.8	0.8	3.0	2	2.4	3 - 5	--	1.5 - 2	82	1,500	--
Lynnwood WWTP	Lynnwood, WA	Jan04-Dec04	4.2	10.9	12 - 23	2.7	0.3	2.4	2	2.5	2	--	2.9	125	1,900	46
Columbia Blvd WWTP	Portland, OR	Sep03-Aug04	65	126	12 - 21	3.0	0.8	2.1	1	2.0	1	0.1 - 0.5	0.1 - 4.8	243	2,000	90
Southeast WPCP	San Francisco, CA	Feb03-Dec03	79	148	18 - 26	1.4	0.4	1.0	2	2.0	4	--	5.0 - 7.5	168	2,400	63
Central Contra Costa Sanitary District	Martinez, CA	Sep03-Aug04	44	82	18 - 27	1.5	0.2	1.3	1	5.9	2	--	0 - 5	140	1,300	--
Oro Loma Sanitary District	San Lorenzo, CA	Jan03-Dec03	14	28	15 - 24	3.3	0.5	2.8	1	1.1	6	--	--	123[4]	1,300	53[4]
Dublin San Ramon Sanitary District	Pleasanton, CA	Jan04-Dec04	12	20	20 - 28	2.3	0.4	1.9	1	2.4/3.0[3]	2 - 3	--	1.3 - 1.9	128	1,900	42
Billings WWTP (Train 2)	Billings, MT	Jan04-Dec04	5.3	8	11 - 20	3.6	1.1	2.6	2	3.4	2	--	1.0 - 1.9	102[5]	2,000	41[5]
Billings WWTP (Train 3)	Billings, MT	Jan04-Dec04	5.3	8	11 - 20	3.0	0.2	2.8	1	1.0	2	--	1.0 - 1.5	102[5]	1,900	35[5]
East Bay Municipal Utility District[2]	Oakland, CA	Jun03-Oct03	34	39	21 - 28	1.3	0.3	1.0	1	1.2	3	--	--	230[6]	2,000	106[6]
Orange County Sanitation District[2]	Fountain Valley, CA	Jul04-Nov04	29	36	27 - 30	1.6	0.3	1.4	1	1.2	5	--	1.0 - 2.5	172	800	149

Plant Name	Selector Loading F/ΣM [kg BOD5/(kg MLSS-d)] S_{x-1}	S_{x-2}	Total	Eff. ICZ	Avg. Selector HRT (w/o recycle) (h) S_{x-1}	S_{x-2}	S_{x-3-6}	Total	Eff. ICZ	Avg. Selector HRT (w/recycle) (h) S_{x-1}	Total	ICZ	Clarifier Underflow Rate (%)	Clarifier Underflow Range (%)	Avg. SVI (mL/g)	90th %ile SVI (mL/g)	% of SVIs > 150 (mL/g)	Avg. Merkel DSVI (mL/g)	90th %ile Merkel DSVI (mL/g)	Dominant Filamentous Organisms
Anoxic																				
Pierce County Chambers Creek WWTP	6.6	3.5	0.87	6.6	0.2	0.2	1.3	1.8	0.2	0.10	0.60	0.10	29%	20-35%	249	310	99%	178	195	1863
Yakima WWTP	0.36	--	--	0.48	2.4	--	--	2.4	1.8	1.42	1.42	1.06	70%	65-75%	121	165	15%	114	142	*S. natans*, 021N, *Thiothrix I/II*
OMI - Plant 6	4.4	2.2	0.74	4.4	0.2	0.2	0.7	1.0	0.2	0.13	0.81	0.13	25%	10-55%	91	127	1%	83	102	021N, *Thiothrix*
Anaerobic																				
King County South TP[1]	6.1	3.5	--	9.1	0.3	0.2	--	0.5	0.2	0.20	0.34	0.13	39%	30-55%	147	212	26%	130	171	1863
OMI - Plant 4	3.6	1.7	0.80	9.3	0.6	0.7	1.5	2.8	0.2	0.39	0.86	0.15	65%	40-80%	127	160	20%	126	156	*M. parvicella*
Tryon Creek WWTP	2.4	1.2	--	2.9	0.6	0.6	--	1.2	0.5	0.47	0.94	0.39	25%	10-45%	189	273	73%	171	214	1701, *N. limicola II*
Lynnwood WWTP	5.6	2.8	--	7.1	0.3	0.3	--	0.6	0.2	0.23	0.45	0.18	35%	25-50%	N/A	N/A	N/A	147[7]	212[7]	1863, 1701 [8]
Columbia Blvd WWTP	2.1	--	--	4.1	1.5	--	--	1.5	0.7	1.11	1.11	0.56	35%	25-45%	170	252	62%	153	203	1701
Southeast WPCP	6.0	2.7	--	6	0.3	0.4	--	0.7	0.3	0.24	0.53	0.24	30%	25-35%	120	156	13%	112	137	No dominant filaments identified.
Central Contra Costa Sanitary District	4.5	--	--	45	0.6	--	--	0.6	0.1	0.42	0.42	0.04	40%	30-45%	139	176	24%	139	176	021N
Oro Loma Sanitary District	2.0[4]	--	--	2[4]	1.2	--	--	1.2	1.0	0.60	0.60	0.52	95%	70-110%	147	205	33%	132	170	No dominant filaments identified.
Dublin San Ramon Sanitary District	2.0	--	--	5.0	1.0	--	--	1.0	0.4	0.58	0.58	0.22	65%	50-80%	100	139	7%	98	136	*Thiothrix*,0914,0041,0675,*H. hydrossis*,*N. limicola*,1863[8]
Billings WWTP (Train 2)	4.4[5]	1.1[5]	--	4.6[5]	0.3	0.9	--	1.2	0.3	0.21	0.85	0.20	45%	40-60%	153	172	48%	144	166	*M. parvicella*
Billings WWTP (Train 3)	4.5[5]	--	--	4.7[5]	0.3	--	--	0.3	0.3	0.19	0.19	0.18	60%	50-70%	222	313	84%	182	239	*M. parvicella*, 1701
East Bay Municipal Utility District[2]	5.1[6]	--	--	5.9[6]	0.6	--	--	0.6	0.5	0.43	0.43	0.37	30%	25-35%	120	166	19%	117	158	021N, *S. natans*, 1701
Orange County Sanitation District[2]	5.7	--	--	6.5	1.0	--	--	1.0	0.8	0.55	0.55	0.48	75%	60-90%	518	731	100%	420	579	021N, *Thiothrix*, *S. natans*, 1701

Notes:

[1] Operation in this mode accounts for 50% of study period.

[2] Data repesents trains with selectors only.

[3] Values correspond to half- and full-compartment anaerobic selector operating modes, respectively.

[4] Reported by facility on a COD basis. BOD_5 estimated as 0.4 x COD based on available plant data.

[5] Reported on a $cBOD_5$ basis. BOD_5 estimated to be equivalent to $cBOD_5$ based on available plant info.

[6] Reported on a $cBOD_5$ basis. BOD_5 estimated as 1.45 x $cBOD_5$ based on available plant data.

[7] Data reported by plants on a DSVI basis. No Merkel equation correction was applied.

[8] Multiple filaments reported, but relative dominance unknown.

Table 4-10. Short-MCRT Plants with Aerobic Selectors.

Plant Name	Location	Study Period	Avg. Flow (MGD)	Peak Flow (MGD)	Temp. Range (°C)	System MCRT (d)	Selector MCRT (d)	Aerated MCRT (d)	No. of Selector Stages	Eff. No. of Selector Stages	No. of Main Aeration Stages	Selector ICZ DO (mg/L)	Aeration Basin DO (mg/L)	Sec. Inf. Avg. BOD (mg/L)	MLSS (mg/L)	Contact Loading (mg BOD/g TSS)
Aerobic																
King County South TP[1]	Renton, WA	Jan03-Aug04	82	154	13 - 23	2.8	0.3	2.8	2	2.6	8	--	2.0 - 2.5	156	1,800	59
Deer Island WWTP Batteries A&B	Boston, MA	Apr04-Mar05	197	333	10 - 25	1.5	0.4	1.5	3	3.0	4	--	8 - 20	83	1,600	49
Deer Island WWTP Battery C	Boston, MA	Apr04-Mar05	135	175	10 - 25	1.4	0.3	1.4	2	2.3	4	--	9 - 20	83	1,500	52

Plant Name	Selector Loading F/ΣM [kg BOD5/(kg MLSS-d)]				Avg. Selector HRT (w/o recycle) (h)					Avg. Selector HRT (w/recycle) (h)			Clarifier Underflow Rate (%)	Clarifier Underflow Range (%)	Avg. SVI (mL/g)	90th %ile SVI (mL/g)	% of SVIs > 150 (mL/g)	Avg. Merkel DSVI (mL/g)	90th %ile Merkel DSVI (mL/g)	Dominant Filamentous Organisms
	S_{x-1}	S_{x-2}	Total	Eff. ICZ	S_{x-1}	S_{x-2}	S_{x-3-6}	Total	Eff. ICZ	S_{x-1}	Total	ICZ								
Aerobic																				
King County South TP[1]	7.2	4.2	--	10.9	0.3	0.2	--	0.5	0.2	0.21	0.36	0.14	40%	25-50%	196	402	50%	164	262	1701, 021N
Deer Island WWTP Batteries A&B	15	7.5	3.8	15	0.1	0.1	0.2	0.4	0.1	0.07	0.29	0.07	30%	25-35%	100	132	5%	99	126	*Thiothrix*, 021N, *S. natans* [2]
Deer Island WWTP Battery C	8.0	4.0	--	9.3	0.2	0.2	--	0.4	0.2	0.14	0.29	0.12	30%	30-50%	116	163	14%	112	158	*Thiothrix, 021N, S. natans* [2]

Notes:

[1] Operation in this mode accounts for 50% of study period.

[2] Multiple filaments reported, but relative dominance unknown.

Table 4-11. Long-MCRT Plants with Selectors.

Plant Name	Location	Study Period	Avg. Flow (MGD)	Peak Flow (MGD)	Temp. Range (°C)	System MCRT (d)	Selector MCRT (d)	Aerated MCRT (d)	No. of Selector Stages	Eff. No. of Selector Stages	No. of Main Aeration Stages	Selector ICZ DO (mg/L)	Aeration Basin DO (mg/L)	Sec. Inf. Avg. BOD (mg/L)	MLSS (mg/L)	Contact Loading (mg BOD/g TSS)
Aerobic																
Winston-Green WWTP[1]	Winston, OR	Jan03-Dec03	1.8	5.2	12 - 20	15	1.2	12	1	1.0	2	--	3.6 - 6.8	141	3,800	48
Upper Occoquan Sewage Authority	Centreville, VA	Jan04-Feb05	27	41	8 - 25	27	0.5	27	1	2.5	2 - 3	0.5 - 1.5	1.3 - 3.3	167[3]	5,900	16[3]
Anoxic																
Springfield Northwest WWTP	Springfield, MO	Jan03-Dec03	3.8	7.4	9 - 25	12	2.1	9.7	1	1.1	2	0.5	0.6 - 5.9	230	2,300	38
Ashland WWTP	Ashland, OR	Nov03-Oct04	2.2	4.7	13 - 22	17	3.4	14	1	1.3	1	0.4	1.6 - 2.7	205[4]	3,500	--
Bend WWTP	Bend, OR	Jan04-Dec04	5.1	6.4	14 - 24	10	3.3	7.0	3	3.0	2	--	2.9 - 3.2	208	2,200	--
Olympus Terrace Sewer District	Mukilteo, WA	Jan04-Dec04	1.8	3.1	12 - 22	24	0.7	24	1	1.4	2	--	0.2 - 2.9	184	3,300	54
Winston-Green WWTP[1]	Winston, OR	Jan03-Dec03	0.9	1.6	15 - 25	16	1.2	14	1	1.0	4	< 0.5	1.6 - 5.6	208	3,900	68
OMI - Plant 1	Southeastern US	Jul03-Jul04	21	39	14 - 26	8.5	2.1	6.4	2	4.0	2	0.2	4.0 - 7.3	186	3,500	43
OMI - Plant 2	Southeastern US	Jul03-Jul04	0.82	2.4	11 - 25	38	7.7	26	3	3.0	3	--	--	130	3,000	27
OMI - Plant 3	Southwestern US	Jan03-Jul04	1.7	2.1	17 - 26	13	3.1	10	3	3.0	1	0.0	1.0 - 4.0	369	2,900	--
OMI - Plant 5	Pacific Northwest	Oct03-Oct04	0.88	3.8	11 - 21	11	3.0	8.1	3	3.5	1	0.3	2 - 3	227	2,700	--
Veolia - Plant 1	Midwestern US	Jan04-Dec04	0.17	0.37	7 - 23	51	26	26	1	2.9	1	0.2	0.1 - 9.7	407	5,100	80
Veolia - Plant 4	New England	Mar04-Mar05	11	17	14 - 29	12	2.4	10	2	3.0	1	--	2.1 - 4.1	183	3,700	28
Veolia - Plant 11	New England	Jan04-Dec04	11	21	10 - 36	21	5.2	16	1	1.2	2	0.1 - 0.3	2.0 - 3.8	246	4,100	28
Veolia - Plant 12	South Central US	Jan04-Dec04	0.4	1.3	18 - 32	21	2.5	19	6	6.0	4	--	1.3 - 6.5	818	3,000	--
Phoenix 91st Avenue WWTP	Tolleson, AZ	Jul03-Jun04	130	164	20 - 32	8.2	1.3	5.7	3	3.2	8	--	1.1 - 2.0	149[6]	3,200	--
Puyallup WPCP	Puyallup, WA	May04-Apr05	3.7	10	12 - 23	24	9.1	10	4	4.0	4	0.2	0.3 - 2.1	170[7]	3,000	--
City of Columbus Southerly WWTP	Lockbourne, OH	Jul03-Jun04	109	221	13 - 25	11	2.7	8.0	4	4.0	6	--	2.7 - 7.8	103	3,400	23
Davenport WPCP	Davenport, IA	Jun03-Nov04	21	41	10 - 25	21	2.0	19	3	3.0	3	0.7	0.5 - 1.0	169[8]	1500[9]	83[8]
Southside WWTP	Tulsa, OK	Jan04-Dec04	31	60	13 - 27	4.1	0.1	3.9	2	2.0	1	0.5	2.2 - 5.1	259	3,100	43
Northside WWTP[2]	Tulsa, OK	Jan04-Dec04	16	38	9 - 26	12	0.7	11	2	3.8	1	0.5	2.9 - 5.2	129	2,100	42
Phoenix 23rd Avenue WWTP	Phoenix, AZ	Jul03-Aug04	48	60	23 - 34	8.1	1.4	6.6	3	3.3	5	--	2.0 - 2.7	160	3,100	--
City of Brewer WPCF	Brewer, ME	Jan04-Dec04	1.9	5.1	5 - 21	11	1.1	9.9	1	1.0	1	0.2 - 1.2	3.2 - 8.6	105	3,600	21
Gilbert Neely WWRF	Gilbert, AZ	Oct03-Sep04	7.6	9.9	24 - 32	13	2.6	10	5	5.9	1	0.2	0.7 - 2.4	178	2,000	14
Glendale WWTF	Lakeland, FL	May04-Apr05	8.9	29	19 - 31	5.3	1.2	4.1	3	3.0	4	0.2 - 0.8	0.7 - 2.8	413	4,300	--

Table 4-11. Long-MCRT Selector Plants (cont'd).

Plant Name	Selector Loading F/ΣM [kg BOD5/(kg MLSS-d)] $S_{x\text{-}1}$	$S_{x\text{-}2}$	Total	Eff. ICZ	Avg. Selector HRT (w/o recycle) (h) $S_{x\text{-}1}$	$S_{x\text{-}2}$	$S_{x\text{-}3\text{-}6}$	Total	Eff. ICZ	Avg. Selector HRT (w/recycle) (h) $S_{x\text{-}1}$	Total	ICZ	Clarifier Underflow Rate (%)	Clarifier Underflow Range (%)	Avg. SVI (mL/g)	90th %ile SVI (mL/g)	% of SVIs > 150 (mL/g)	Avg. Merkel DSVI (mL/g)	90th %ile Merkel DSVI (mL/g)	Dominant Filamentous Organisms
Aerobic																				
Winston-Green WWTP[1]	1.8	--	--	1.8	0.6	--	--	0.6	0.6	0.36	0.36	0.36	55%	30-80%	128	201	36%	97	129	*N. limicola*
UOSA	2.5[3]	--	--	6.3 [3]	0.2	--	--	0.2	0.1	0.14	0.14	0.05	75%	40-140%	96	155	12%	66	92	*M. parvicella*
Anoxic																				
Springfield Northwest WWTP	0.39	--	--	0.44	6.5	--	--	6.5	5.8	2.38	2.38	2.12	175%	125-200%	129	154	13%	124	146	*N. limicola, S. natans, M. parvicella* [11]
Ashland WWTP	0.18[4]	--	--	0.24 [4]	7.8	--	--	7.8	5.9	3.99	3.99	3.02	--	35-40%	89	128	4%	81	103	0675, 0041
Bend WWTP	2.2	1.1	0.55	2.2	1.1	1.1	2.2	4.4	1.1	0.23	0.93	0.23	50%	--	243	441	77%	168	224	*M. parvicella*
Olympus Terrace Sewer District	2.6	--	--	3.6	0.6	--	--	0.6	0.4	0.40	0.40	0.28	45%	20-80%	112	132	2%	101	114	0675, 0041
Winston-Green WWTP[1]	1.4	--	--	1.4	1.0	--	--	1.0	1.0	0.61	0.61	0.61	55%	40-70%	77	104	2%	74	96	*N. limicola*
OMI - Plant 1	1.0	0.50	--	2.0	1.3	1.3	--	2.6	0.7	0.94	1.89	0.48	40%	30-50%	98	122	2%	87	102	0675
OMI - Plant 2	0.58	0.29	0.14	0.58	1.9	1.9	4.0	7.8	1.9	1.17	4.80	1.17	65%	40-90%	187	239	80%	128	146	0675, *H. hydrossis*
OMI - Plant 3	1.3	0.65	0.43	1.3	2.3	2.3	2.3	6.9	2.3	0.27	0.82	0.27	35%	30-45%	127	195	29%	105	132	1851, *M. parvicella*
OMI - Plant 5	1.3	0.64	0.43	1.5	1.7	1.7	1.7	5.1	1.5	0.47	1.423	0.41	30%	20-50%	108	149	9%	107	144	1851
Veolia - Plant 1	0.095	--	--	0.22	22	--	--	22	7.8	10.11	10.11	3.55	115%	--	174	202	91%	92	107	*M. parvicella* , 0041, 0675, 1851
Veolia - Plant 4	1.3	0.63	--	1.9	1.0	1.0	--	2.0	0.7	0.52	1.04	0.35	85%	60-110%	119	156	12%	94	105	No dominant filaments identified.
Veolia - Plant 11	0.59	--	--	0.68	2.6	--	--	2.6	2.3	0.52	0.52	0.45	110%	80-160%	93	140	8%	76	95	0041, 0675
Veolia - Plant 12	4.3	2.1	0.53	4.3	1.9	1.9	11	15	1.9	0.71	5.71	0.71	140%	--	131	180	20%	110	138	0675, *N. limicola I,* 0041
Phoenix 91st Avenue WWTP	3.9[6]	2.0[6]	1.0[6]	3.9[6]	0.3	0.3	0.6	1.1	0.3	0.07	0.26	0.07	50%	50-55%	N/A	N/A	N/A	71[10]	76[10]	*Thiothrix II* , 0092, 0675, 1701, 0041
Puyallup WPCP	2.2[7]	1.1[7]	0.16[7]	2.2[7]	0.7	0.7	8.1	9.5	0.7	0.34	2.31	0.34	15%	10-25%	106	122	0%	99	115	0041
City of Columbus Southerly	3.4	1.7	0.38	3.4	0.3	0.3	1.8	2.4	0.3	0.18	1.58	0.18	50%	N/A	100	114	0%	88	100	021N
Davenport WPCP	4.9[8]	2.5[8]	1.6[8]	4.9[8]	0.6	0.6	0.6	1.9	0.6	0.47	1.40	0.47	35%	20-50%	128	235	18%	105	166	021N, *N. limicola I/II*
Southside WWTP	21	10	--	21	0.1	0.1	--	0.2	0.1	0.06	0.11	0.06	75%	55-105%	101	161	12%	90	114	No dominant filaments identified.
Northside WWTP[2]	5.5	2.8	--	11	0.3	0.3	--	0.6	0.1	0.18	0.36	0.09	60%	40-80%	114	138	4%	113	136	0581
Phoenix 23rd Avenue WWTP	6.6	2.5	1.0	6.6	0.2	0.3	0.7	1.2	0.2	0.06	0.31	0.06	50%	45-50%	N/A	N/A	N/A	81[10]	101[10]	1701, *Thiothrix II* , 1851, 0675, 0041, 0914
City of Brewer WPCF	1.0	--	--	1.1	0.7	--	--	0.7	0.7	0.46	0.46	0.44	55%	35-80%	75	102	0%	74	95	*N. limicola I*
Gilbert Neely WWRF	4.8	2.4	0.4	4.8	0.4	0.4	4.4	5.3	0.4	0.07	0.86	0.07	160%	110-200%	116	184	17%	112	166	0092, *M. parvicella* [11]
Glendale WWTF	4.4	2.2	1.2	4.4	0.6	0.6	0.9	2.0	0.6	0.16	0.43	0.16	60%	30-120%	96	180	16%	77	107	*N. limicola II/III*

Notes:

[1] Operation in this mode accounts for 50% of study period.

[2] Data repesents trains with selectors only.

[3] Reported on a COD basis. BOD_5 estimated as 0.5 x COD loading.

[4] Reported by facility on a $cBOD_5$ basis. BOD_5 estimated as 1.1 x $cBOD_5$ based on available plant data.

[5] Reported by facility as plant influent BOD_5. Secondary influent BOD_5 estimated as 0.7 x plant influent BOD_5.

[6] Reported by facility on a COD basis. BOD_5 estimated as 0.46 x COD based on available plant data.

[7] Reported by facility on a COD basis. BOD_5 estimated as 0.45 x COD based on available plant data.

[8] Reported on a $cBOD_5$ basis. BOD_5 estimated to be equivalent to $cBOD_5$ based on available plant information.

[9] Reported value is from contact zone of contact-stabilization plant. Stabilitization zone suspended solids concentrations is 6,000 mg/L.

[10] Data reported by plants on a DSVI basis. No Merkel equation correction was applied.

[11] Multiple filaments reported, but relative dominance unknown.

Tables 4-12, 4-13, and 4-14 show results of the single-regression analyses for short-MCRT plants with anoxic or anaerobic selectors, short-MCRT plants with aerobic selectors, and long-MCRT plants with selectors, respectively. The parameters are listed in order of their relative influence on DSVI, as measured by R^2 values. Please note that R^2 values will be lower for single-regression analyses compared to multiple regression analyses when more than one independent variable has influence on the dependent variable. Although the R^2 values in Tables 4-12 through 4-14 can be very low, because the number of samples are large in these analyses (approximately 1,000 to 9,000 samples for each independent variable), the R^2 value can still be used to provide a ranking of relative influence of the design/operating parameters on DSVI. For further discussion on R^2 and this study's regression analyses, please see Appendix D.

Table 4-12. Short-MCRT Plants with Anoxic or Anaerobic Selectors: Significant Parameters.

Parameter	T-Statistic	R^2 (%)
Average MLSS (mg/L)	-39.51	22.4
7-d Avg. Reactor MCRT (d)	-26.41	12.4
Selector F/M [kg BOD_5/(kg MLSS·d)]	26.03	11.9
7-d Avg. Selector MCRT (d)	-22.23	9.1
7-d Avg. Selector MCRT >2 d	-21.12	8.3
Number of Selector Stages	-20.77	7.7
Aeration Basin DO (mg/L)	-16.78	7.3
7-d Avg. Selector MCRT >1 d	-18.93	6.8
Effective No. of Selector Stages[1]	-18.67	6.4
	cubic polynomial[2]	7.3
7-d Avg. Selector MCRT >3 d	-15.60	4.7
Activ. Sludge Influent BOD_5/TSS Ratio	14.56	4.1
ICZ F/M [kg BOD_5/(kg MLSS·d)]	12.82	3.2
Nominal Selector HRT (without recycle) (h)	-10.53	2.1
Selector Vol./Total Basin Vol. Ratio	-10.30	2.0
Selector HRT (with recycle) (h)	-9.85	2.0
ICZ HRT (with recycle) (h)	-8.87	1.6
Nominal ICZ HRT (without recycle) (h)	-7.83	1.2
Effluent Temperature (°C)	-7.63	1.1
Number of Aeration Basin Stages	-5.45	0.6
Activ. Sludge Inf. BOD (mg/L)	3.36	0.2
BOD_5 Loading (lbs BOD_5/d)	-2.50	0.1
Effective ICZ HRT (with recycle) (h)[1]	-1.38	0.1
	cubic polynomial[2]	5.5
Effective ICZ F/M [kg BOD_5/(kg MLSS·d)] [1]	2.16	0.1
	cubic polynomial[2]	1.9
% RAS Flow (%)	1.98	0.1
Effluent pH	1.81	0.0
Effective Nominal ICZ HRT (without recycle) (h)[1]	-2.35	0.0
	cubic polynomial[2]	6.9

Approx. No. of Samples: 5,150

Notes: [1] Based on calculated number of stages (N) – see Section 4.3.2.3
[2] Cubic polynomial regression analysis R^2 for previous parameter

Table 4-12 suggests that higher MLSS (R^2 = 22.4%) concentrations, longer total reactor MCRT (R^2 = 12.4%), lower total selector F/M (R^2 = 11.9%), and longer selector MCRT (R^2 = 9.1%) will reduce DSVI in short-MCRT systems with unaerated selectors. In contrast, the higher the ICZ F/M the higher the DSVI, which is probably a result of the very strong correlation between ICZ F/M and total selector F/M. This is supported by the weaker influence the ICZ F/M has on DSVI (R^2 = 3.2%) compared to that of the total selector F/M (R^2 = 11.9%). Because the ICZ F/M is positively correlated to DSVI (increasing ICZ F/M does not result in decreased DSVI, as will be shown with the short-MCRT plants with aerobic selectors), this suggests that kinetics play no significant role in the ability of an unaerated selector to control filamentous bulking in short-MCRT systems. However, increasing the number of selector stages (R^2 = 7.7%), increasing the selector HRT (R^2 = 2.1%), or increasing the selector volume to total basin volume ratio (R^2 = 2.0%) does correlate with lower DSVIs.

This correlation suggests that the anaerobic/anoxic selector basins should be sized large enough to remove all or most of the raCOD[1] and staged to prevent short-circuiting and raCOD breakthrough to the main aeration basin rather than to provide a better kinetic advantage (more rapid raCOD uptake). This is consistent with Wanner (1994), where a single-stage anaerobic selector controlled filamentous bulking, and an increase in the selector volume to total basin volume ratio significantly reduced SVIs (from 800 mL/g to less than 100 mL/g) when short-MCRT filamentous organisms were initially dominant. Based on these results, well-designed and well-operated unaerated selector systems appear to be effective in controlling filamentous organisms and bulking in short-MCRT systems. In addition to selector design and operating parameters, higher aeration basin DO (R^2 = 7.3%) and lower secondary influent BOD/TSS ratios (R^2 = 4.1%) corresponded to lower DSVIs in short-MCRT WWTPs.

Design and operating parameters for short-MCRT plants with aerobic selectors are listed in Table 4-13. This table shows that the parameters with the most influence on DSVI are the activated sludge influent BOD_5 concentration (R^2 = 36.7%), ICZ HRT (R^2 = 33.7%), selector HRT (R^2 = 25.7%), and %RAS flow (R^2 = 21.0%). When any of these parameters are increased, the DSVI increases. On the other hand, when the ICZ F/M is increased (which corresponds to decreasing the ICZ HRT), the DSVI decreases. This supports the hypothesis that aerated selectors require a BOD_5 concentration gradient to favor floc-forming bacteria over filamentous bacteria and control bulking; but if the influent BOD_5 concentration is too high, raCOD may leak through the selector allowing filamentous organism growth and bulking.

Table 4-13. Short-MCRT Plants with Aerobic Selectors: Significant Parameters.

Parameter	T-Statistic	R^2 (%)
Activ. Sludge Inf. BOD_5 (mg/L)	24.29	36.7
Nominal ICZ HRT (without recycle) (h)	22.81	33.7
ICZ HRT (with recycle) (h)	21.33	30.8
Effluent pH	20.41	29.4
Nominal Selector HRT(without recycle) (h)	18.79	25.7
% RAS Flow (%)	16.47	21.0
Selector HRT (with recycle) (h)	16.02	20.1
Effluent Temperature (°C)	14.70	17.4
ICZ F/M [kg BOD_5/(kg MLSS·d)]	-10.28	9.4
7-d Avg. Reactor MCRT (d)	7.84	5.8
Average MLSS (mg/L)	7.12	4.7
Aeration Basin DO (mg/L)	-5.43	3.9
BOD Loading (lbs BOD_5/d)	-5.43	2.8
Activ. Sludge Influent BOD_5/TSS Ratio	4.73	2.2
Selector F/M [kg BOD_5/(kg MLSS·d)]	4.22	1.7
7-d Avg. Selector MCRT (d)	-4.10	1.6
7-d Avg. Selector MCRT >1 d	-1.41	0.2

Approx. Number of Samples: 1,020

[1] Refer to raCOD discussion on Page 1-3 in Chapter 1.0.

Table 4-14 lists selector design and operating parameters in order of their relative influence on DSVI (measured with R^2 values) for the long-MCRT WWTPs studied. The parameters that appear to have the most dominant influence on DSVI in the long-MCRT WWTPs are: 1) MLSS (the higher the MLSS, the lower the DSVI), 2) selector HRT (the lower the HRT, the lower the DSVI, with an HRT = 0 being the best), and 3) ICZ HRT (with an ICZ HRT → 0 providing the best DSVI). None of these dominant parameters suggests that selectors have much influence on DSVI in longer MCRT WWTPs. In fact, larger selector volume to total basin volume ratios correspond to higher DSVIs, the opposite of what was found for short-MCRT WWTPs.

Table 4-14. Long-MCRT Plants with Selectors: Significant Parameters.

Parameter	T-Statistic	R^2 (%)
Average MLSS (mg/L)	-52.69	23.4
Selector HRT (with RAS) (h)	30.55	11.5
ICZ HRT (with RAS) (h)	29.28	10.6
Selector Vol./Total Basin Vol. Ratio	21.36	5.9
BOD_5 Loading (lbs BOD_5/d)	-22.97	5.8
7-d Avg. Selector MCRT >2 d	21.04	4.8
No. of Aeration Basin Stages	-21.45	4.8
Nominal Selector HRT (without recycle) (h)	20.47	4.5
No. of Selector Stages	19.34	4.0
Effluent pH	17.35	3.5
Effective ICZ HRT (with recycle) (h)[1]	15.05	3.2
	cubic polynomial[2]	3.4
Effluent Temperature (°C)	-15.56	2.8
Effective Number of Selector Stages[1]	13.05	2.3
	cubic polynomial[2]	6.4
Effective Nominal ICZ HRT (without recycle)[1]	8.95	1.1
	cubic polynomial[2]	7.4
% RAS Flow (%)	8.92	1.1
Activ. Sludge Inf. BOD_5 (mg/L)	8.99	0.9
7-d Avg. Selector MCRT >3 d	8.74	0.9
Aeration Basin DO (mg/L)	8.10	0.8
Effective ICZ F/M [kg BOD_5/(kg MLSS·d)] [1]	-5.09	0.4
	cubic polynomial[2]	0.9
7-d Avg. Selector MCRT > 1 d	5.49	0.3
Nominal ICZ HRT (without recycle) (h)	4.76	0.3
7-d Avg. Reactor MCRT (d)	-4.83	0.3
Selector F/M [kg BOD_5/(kg MLSS·d)]	-4.26	0.2
Activ. Sludge BOD_5/TSS Ratio	2.85	0.1
7-d Avg. Selector MCRT (d)	2.49	0.1
ICZ F/M [kg BOD_5/(kg MLSS·d)]	-3.31	0.1

Approx. Number of Samples: 9,000

Notes:

[1] Based on calculated number of stages (N) – see Section 4.3.2.3.

[2] Cubic polynomial regression analysis R^2 for previous parameter

Other significant parameters corresponding to lower DSVIs include longer 7-d average total basin (or reactor) MCRTs, larger number of aeration basin stages, smaller number of selector stages, lower effluent pH, and higher effluent temperature. In contrast, selector F/M, selector MCRT, and ICZ F/M had little influence on DSVI (R^2 = 0.2%, 0.1%, and 0.1%, respectively). These results also suggest that selectors may not be effective for controlling filamentous bulking in long-MCRT WWTPs, especially when compared to the selector effect in short-MCRT WWTPs. Therefore, the regression results suggest that short-MCRT filamentous organisms are better controlled with selectors than long-MCRT filamentous organisms, and this is supported by the literature (Wanner, 1994; Jenkins et. al., 2004, etc.).

The design and operating parameters will be discussed in more detail in the following sections for each of the three wastewater treatment plant groups—short-MCRT with unaerated selectors, short-MCRT with aerated selectors, and long-MCRT plants. This discussion will also include the "effective" ICZ parameters and the "effective" number of selector stages, which were determined using the calculated N value estimate for the number of selector stages (see Section 4.3.2.3), rather than the physical number of selector compartments.

Figures 4-18 through 4-42 present plots of selector design and operating parameters versus log DSVI (log DSVI in general correlated slightly better with the parameters tested than did DSVI or SVI) from the same data used in the regression analysis and again separated into short- and long-MCRT WWTPs. Each long-MCRT graph includes approximately 9,000 data points, the short MCRT with unaerated selectors graphs include about 5,000 data points each, and each short MCRT with aerated selectors graph includes slightly over 1,000 data points in many cases. The figures are discussed further in the following sections, which are divided into short- and long-MCRT regression analyses.

4.4.11.1 Short-MCRT Plants with Anoxic or Anaerobic Selectors: Regression Results

Average MLSS

The average MLSS shows the highest correlation with log DSVI, with the highest R^2 value at 22.4% (t-statistic = -39.51), compared to all other parameters tested for this plant group; log DSVI decreases with increasing MLSS (Figure 4-18a). Is this relationship due to the DSVI calculation, or does MLSS influence the DSVI through other parameters? The total selector F/M and the 7-d average reactor MCRT also significantly influence the DSVI (R^2 = 11.9% and 12.4%, respectively), and both are calculated using the MLSS. Regressing MLSS against the total selector F/M gives an R^2 = 15.9% and a t-statistic = -30.81. Regressing MLSS against the 7-d average reactor MCRT gives an R^2 = 33.9% and a t-statistic = 50.47. This demonstrates that MLSS is more strongly correlated to the 7-d average reactor MCRT than to the log DSVI. Therefore, it may be that the MLSS is highly correlated to log DSVI in large part because the MLSS is even more correlated to the reactor MCRT, which is correlated to log DSVI. Further, MLSS is significantly more correlated to DSVI than reactor MCRT is correlated to DSVI (the MLSS R^2 value is almost twice the reactor MCRT R^2 value when both are regressed against DSVI). Nonetheless, the higher MLSS may favor enhanced raCOD uptake in unaerated selectors operated in short-MCRT systems; and by reducing the amount of raCOD leaking into the main aeration basin, less filamentous organisms grow, and the DSVI is reduced.

Figure 4-18a shows the approximately 5,000 data points plotted with the cubic polynomial regression line. Using the regression equation, the regression curve was plotted in Excel for better clarity (Figure 4-18b). Slope lines were drawn on the curve in Figure 4-18b, to demonstrate how the slope of the curve becomes more flattened with increasing MLSS. From these slope lines, best operating MLSS values appears to be between 1,500–2,000 mg/L. The DSVI continues to improve above 2,000 mg/L, but at a much lower rate.

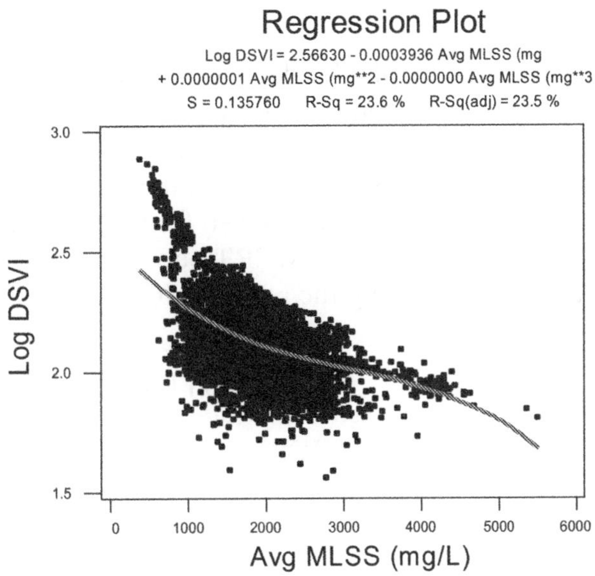

Figure 4-18a. MLSS vs. Log DSVI - Regression Plot – Short-MCRT Plants with Anoxic or Anaerobic Selectors.

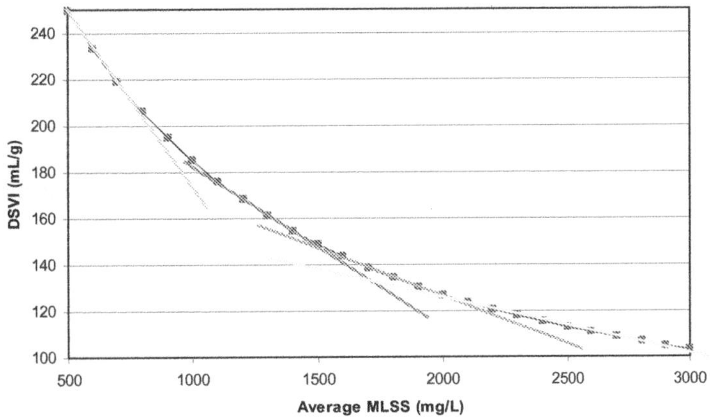

Figure 4-18b. MLSS vs. DSVI - Cubic Polynomial Regression Curve – Short-MCRT Plants with Anoxic or Anaerobic Selectors.

7-day Reactor MCRT

The regression analysis shows that DSVI decreases with increasing reactor MCRT (R^2 = 12.4%, t-statistic = -26.41), and is the second most influential parameter on DSVI in this group. Figure 4-19a shows the polynomial regression line superimposed onto the plotted data points. Most of the data points occur at MCRTs ≤4.5 d; however, there are data points that spread up to 20 d. The regression curve was replotted in Figure 4-19b for better clarity. This curve shows that the DSVI is not substantially affected by reactor MCRT between 0.5–4.5 d, where most of the data points in this group occur, but increasing the reactor MCRT >4.5 d correlates to decreasing DSVI.

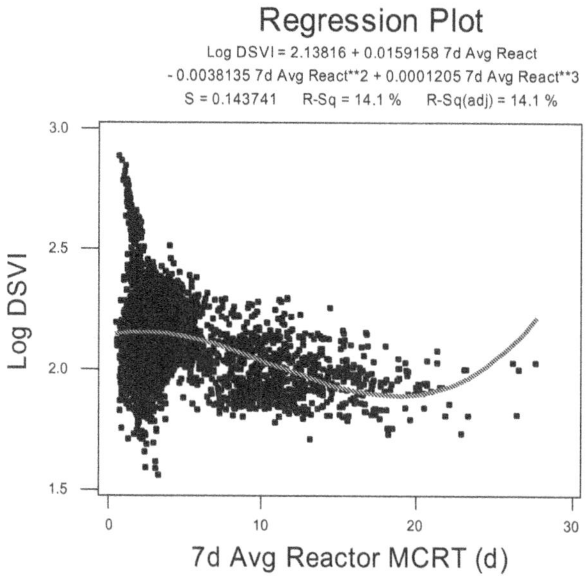

Figure 4-19a. 7-d Average Reactor MCRT vs. Log DSVI - Regression Plot – Short-MCRT Plants with Anoxic or Anaerobic Selectors.

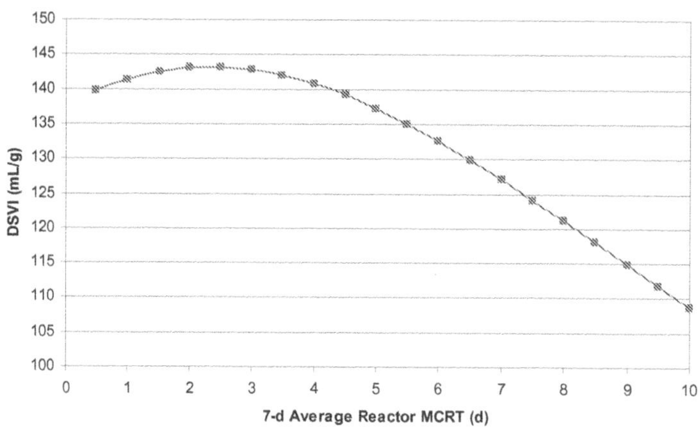

Figure 4-19b. 7-d Average Reactor MCRT vs. DSVI - Cubic Polynomial Regression Curve – Short-MCRT Plants with Anoxic or Anaerobic Selectors.

Total Selector F/M

The total selector F/M is the next most correlated variable with log DSVI (R^2 = 11.9%); the higher the selector F/M the higher the DSVI (Figure 4-20a). The higher the selector F/M, the higher is the likelihood of raCOD leakage into the main aeration basin, which would likely support filamentous growth in activated sludge, resulting in a higher DSVI. Figure 4-20a shows the cubic polynomial regression plot of the selector F/M to log DSVI (R^2 = 14.1%). Figure 4-20b shows that the lower the selector F/M, the lower the DSVI. The selector F/M should be <1.0 kg BOD_5/(kg MLSS·d), if possible, which is consistent with Jenkins et al., 2004. In fact, the slope of the curve in Figure 4-18b is steepest for selector F/Ms from 0.1 to 1.0 kg BOD_5/(kg MLSS·d), so the improvements in DSVI should occur faster as the selector F/M is lowered from 1.0 to 0.1 kg BOD_5/(kg MLSS·d), compared to lowering the selector F/M down to 1.0 kg BOD_5/(kg MLSS·d).

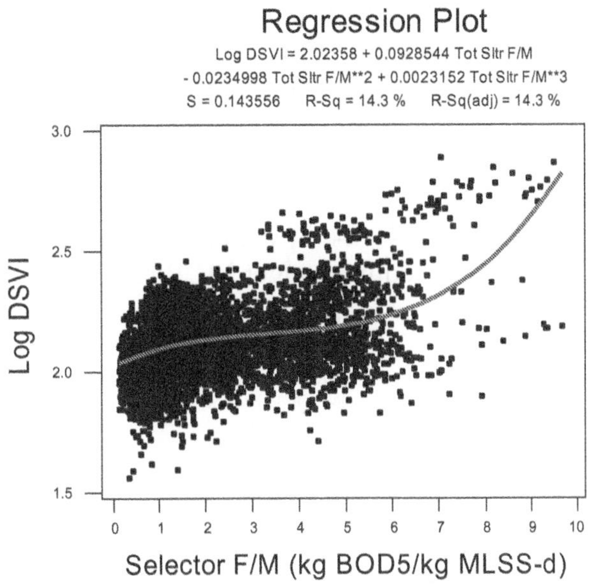

Figure 4-20a. Selector F/M vs. Log DSVI - Regression Plot – Short-MCRT Plants with Anoxic or Anaerobic Selectors.

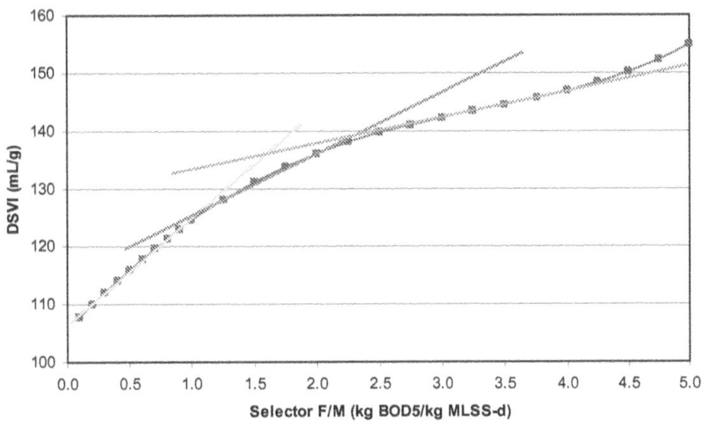

Figure 4-20b. Selector F/M vs. DSVI - Cubic Polynomial Regression Curve – Short-MCRT Plants with Anoxic or Anaerobic Selectors.

The ICZ F/M (R^2 = 3.2%) did not influence DSVI in the short-MCRT plants with anoxic or anaerobic selectors as the total selector F/M did in this study, but the ICZ F/M appears to correlate with the total selector F/M in that the DSVI also increases when the ICZ F/M increases. This suggests that a BOD_5 concentration gradient through the selector, and therefore selector kinetics, is not important to the success of anaerobic or anoxic selectors. Further, DSVI is lowest when the ICZ HRT is about the same as the selector HRT (discussed later in this section).

7-day Selector MCRT

The DSVI decreases with increasing selector MCRT for anoxic and anaerobic selectors, according to the linear regression analysis (R^2 = 9.6%). Considering the polynomial regression line (R^2 = 10.2%, Figure 4-21a), the most improvement (steepest slope) in DSVI occurs as the selector MCRT increases from 0.7 to 2.25 d. Further DSVI reduction occurs at a lower rate as the selector MCRT increases from 2.0 to 3.0 d, but there appears to be much less, if any, DSVI improvement beyond a selector MCRT of 3.0–3.5 d in anaerobic or anoxic selectors (Figure 4-21b). Therefore, ideally, the selector MCRT should be maintained between approximately 2.0–3.0 d, per the regression analysis in this study. This selector MCRT range is in agreement with or somewhat more conservative than Marten and Daigger (1997). Since best actual MCRTs are dependent on the growth rates of preferred organisms in activated sludge (and these growth rates vary with temperature) the actual best selector MCRT likely varies with wastewater temperature. The wastewater temperature varied between 10°–30°C in the short-MCRT regression group, and most temperatures were between about 14°–24°C, with both the mean and median temperature at 18.9°C. In general, lower temperatures require longer MCRTs.

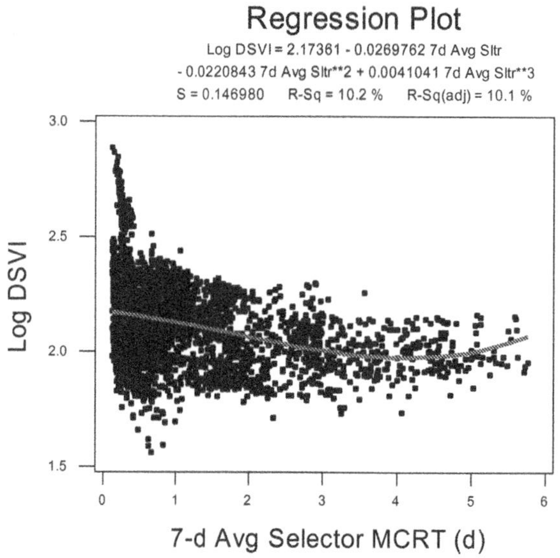

Figure 4-21a. 7-d Average Selector MCRT vs. Log DSVI - Regression Plot – Short-MCRT Plants with Anoxic or Anaerobic Selectors.

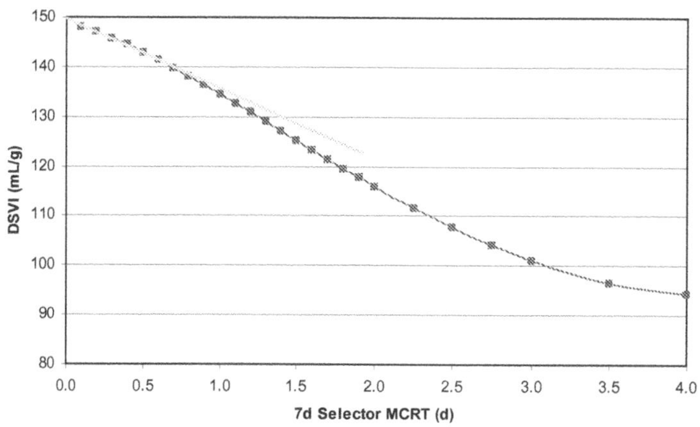

Figure 4-21b. 7-d Selector MCRT vs. DSVI - Cubic Polynomial Regression Curve – Short-MCRT Plants with Anoxic or Anaerobic Selectors.

Number of Selector Stages

According to the regression analysis in this study, staged selectors control filamentous bulking better than unstaged selectors in short-MCRT systems with anoxic or anaerobic selectors, and more stages result in better settleability (R^2 = 7.7%, t-statistic = -20.77, where the database included selectors with one, two, four, and six stages, refer to Figure 4-22a). The cubic polynomial regression curve, however, provided puzzling results (Figures 4-22b and 4-22c). The DSVI rises slightly as the number of selector stages increases from one to three or four, and then drops sharply at five and six stages (R^2 = 15.1%). Removing the one six-stage plant from the database, however, results in the R^2 dropping to 0.4%, even for the cubic multiple regression analysis (Figure 4-22d), and since the new t-statistic is positive (4.22), the DSVI will increase when the number of selector stages increases.

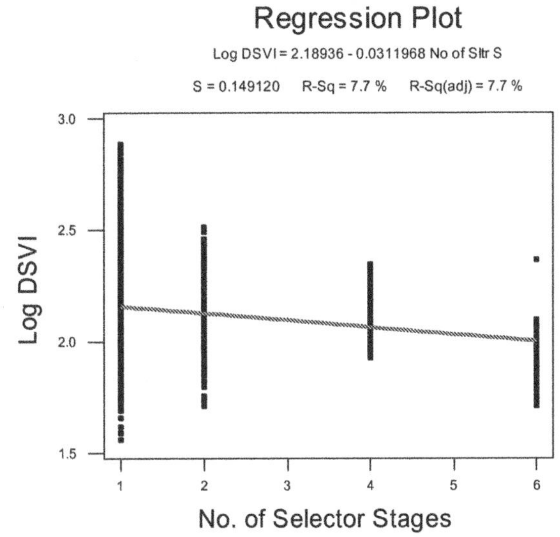

Figure 4-22a. Number of Selector Stages vs. Log DSVI - Linear Regression Plot – Short-MCRT Plants with Anoxic or Anaerobic Selectors.

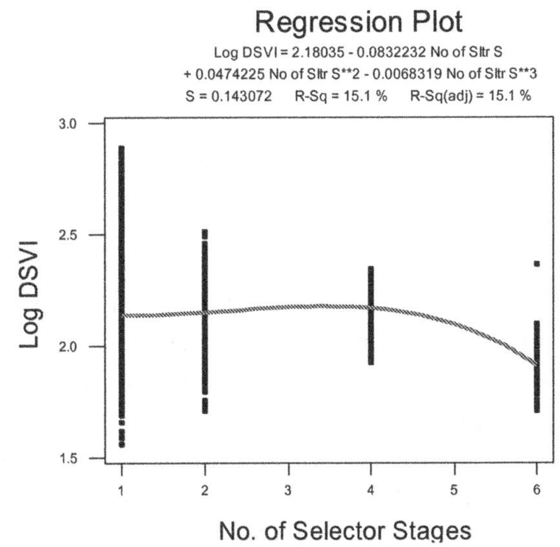

Figure 4-22b. Number of Selector Stages vs. Log DSVI - Cubic Polynomial Regression Plot – Short-MCRT Plants with Anoxic or Anaerobic Selectors.

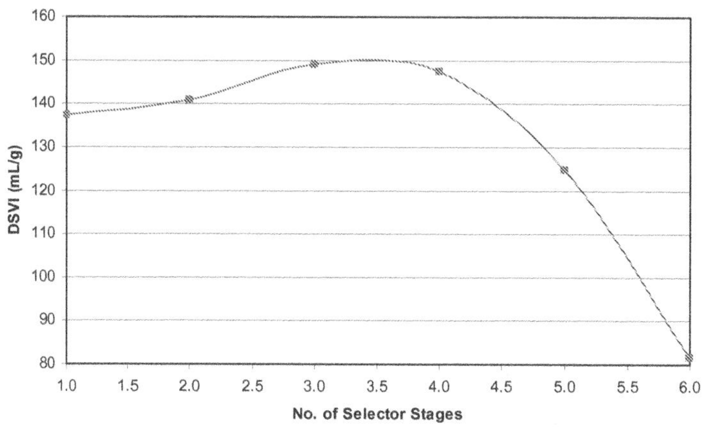

Figure 4-22c. Number of Selector Stages vs. DSVI - Cubic Polynomial Regression Curve – Short-MCRT Plants with Anoxic or Anaerobic Selectors.

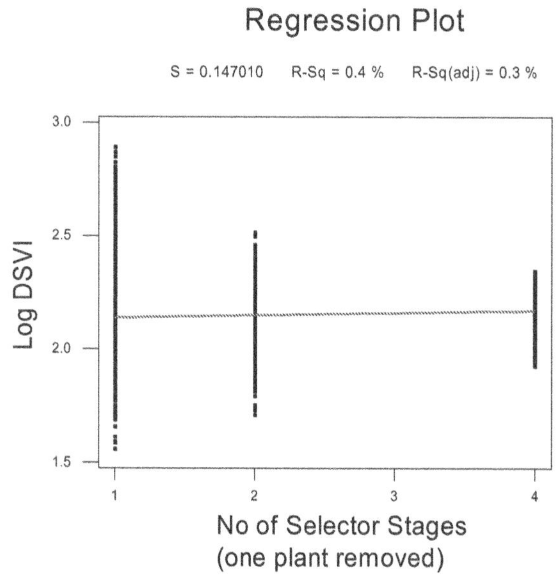

Figure 4-22d. Number of Selector Stages (one plant removed) vs. Log DSVI - Linear Regression Plot – Short-MCRT Plants with Anoxic or Anaerobic Selectors.

The number of stages was counted based on the number of physical compartments built into the selector. The selector length-to-width (L:W) ratios, however, varied widely with the different WWTPs included in this study. This meant that both a cubic structure with all sides equal and a very long narrow channel with a L:W ratio equal to 30 were each counted as single-stage selectors. Using the N equation derived in this study to approximate the equivalent number of selector stages for each plant (see Section 4.3.2.3) resulted in some of the plants having more selector stages, while others were unchanged. The N value provided an "effective" number of selector stages; and since the N value is usually not an integer, the data set was better distributed over more selector stages' values (Figure 4-22e). The cubic polynomial regression R^2 value is 7.3 for the effective number of selector stages, and the regression curve is redrawn in Figure 4-22f for clarity. Figure 4-22f suggests that two selector stages yield as good a result as can be obtained unless the number of selector stages is increased to six.

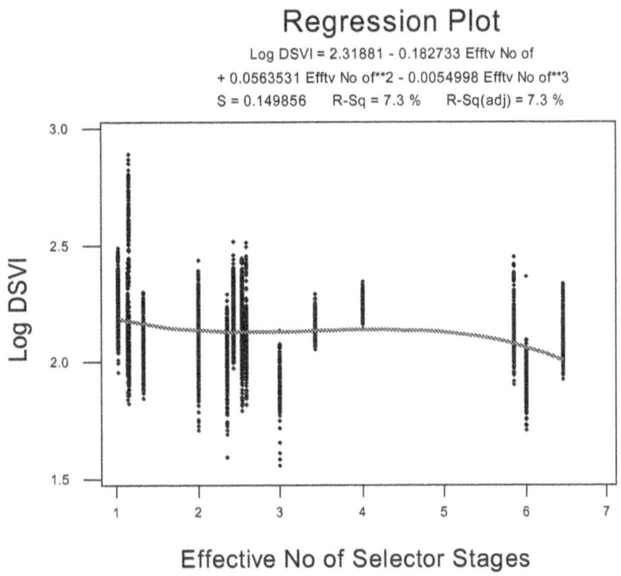

Figure 4-22e. Number of Effective Selector Stages vs. Log DSVI - Cubic Polynomial Regression Plot – Short-MCRT Plants with Anoxic or Anaerobic Selectors.

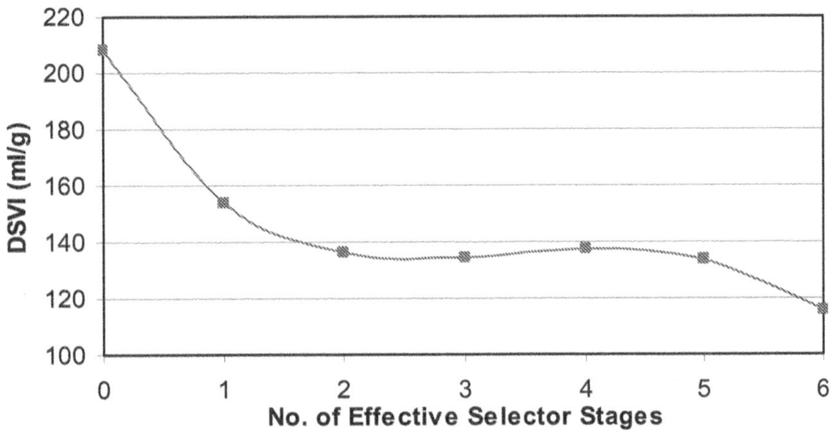

Figure 4-22f. Number of Effective Selector Stages vs. DSVI - Cubic Polynomial Regression Curve – Short-MCRT Plants with Anoxic or Anaerobic Selectors.

The one six-stage plant that was removed from the data set previously may have still been imposing an unusual influence and was therefore removed again. Figure 4-22g shows that the cubic polynomial regression R^2 drops to 3.6% when the one plant is removed. Although the effective number of selector stages is not as influential on DSVI without the plant data set in question, the parameter is still significant. Figure 4-22h shows that DSVI still drops as the number of selector stages is increased to 2.0–2.5, but then the DSVI climbs when more selector stages are added. This may be influenced by one or two other plants with 3.5–4.0 selector stages. Taking these data out would likely result in a relatively flat line from two to six selector stages.

Since DSVI drops when the number of selector stages increases to two, this could suggest that kinetics are important to the success of anaerobic and anoxic selectors; however, as previously discussed, the ICZ F/M is not significant, and the lowest DSVIs occur when the ICZ HRT is equal to the selector HRT, so the importance and benefit of staging is probably not to induce an raCOD concentration gradient. Another benefit to staging is reducing short-circuiting or raCOD bleed-through. This is consistent with the importance of having a lower total selector

F/M to achieve better activated sludge settleability (per the previous selector F/M discussion). This hypothesis is also supported by the result that DSVI is reduced as the selector stages are increased to two but not much more as additional stages are added.

Therefore, it appears that selector staging may be important to reducing DSVI because it reduces short-circuiting and prevents raCOD breakthrough rather than enhancing raCOD uptake kinetics in the selector. The literature suggests a three-stage selector be used in anoxic and anaerobic selector systems (Wanner, 1993; Jenkins, 2004).

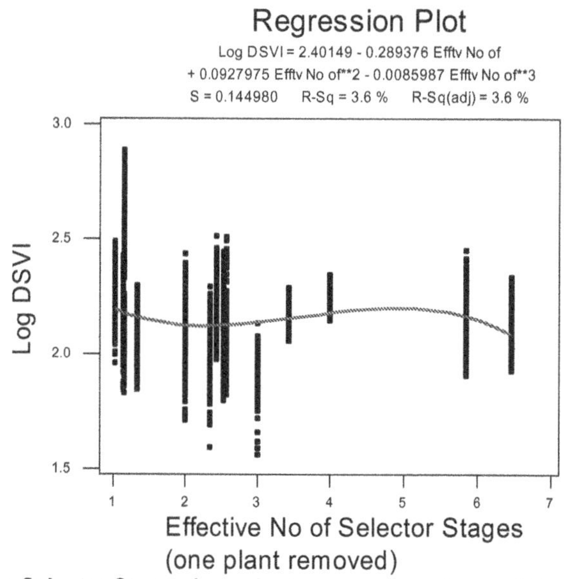

Figure 4-22g. Number of Effective Selector Stages (one plant removed) vs. Log DSVI - Cubic Polynomial Regression Plot – Short-MCRT Plants with Anoxic or Anaerobic Selectors.

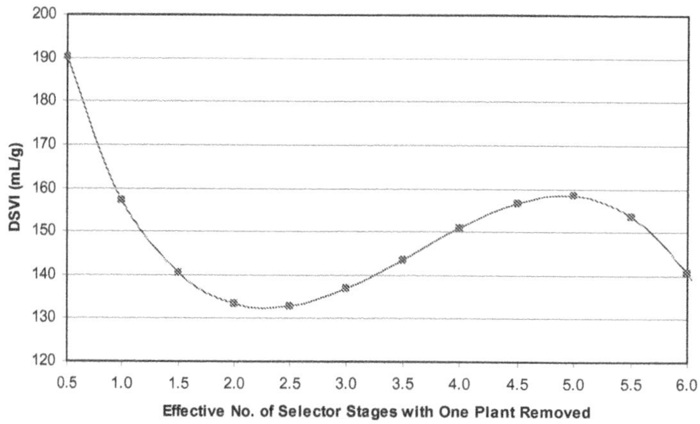

Figure 4-22h. Number of Effective Selector Stages (one plant removed) vs. DSVI - Cubic Polynomial Regression Curve – Short-MCRT Plants with Anoxic or Anaerobic Selectors.

Aeration Basin DO

Aeration basin DO is significantly correlated to log DSVI ($R^2 = 7.4\%$); the higher the aeration basin DO, the lower the DSVI. This is consistent with the well-known observation that many filamentous organisms proliferate in activated sludges with low DO. Since all the WWTPs in this study had selectors, it is possible that a selector may not be as effective under low DO conditions. Further, according to the polynomial regression line ($R^2 = 8.3\%$), the most improvement (steepest slope) occurs as the DO increases from 0 to about 1.5 mg/L, with little added benefit after the DO is increased past 5 mg/L (Figures 4-23a and 4-23b). The best operating range is taken from Figure 4-23b to be about 2.5–4.0 mg/L for DSVI control. This DO concentration range, however, should not be applied to pure oxygen activated sludge plants, since "low DO" filamentous organisms have been observed to thrive at mixed liquor DO concentrations much higher than 5 mg/L (Wanner, 1993; Jenkins et al., 2004).

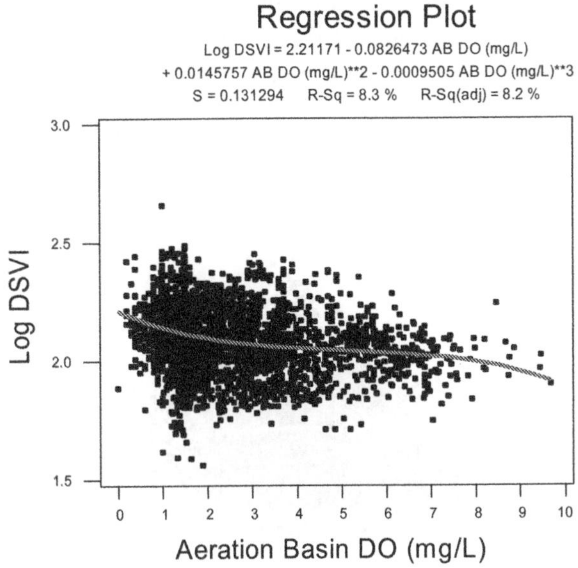

Figure 4-23a. Aeration Basin DO vs. Log DSVI - Cubic Polynomial Regression Plot – Short-MCRT Plants with Anoxic or Anaerobic Selectors.

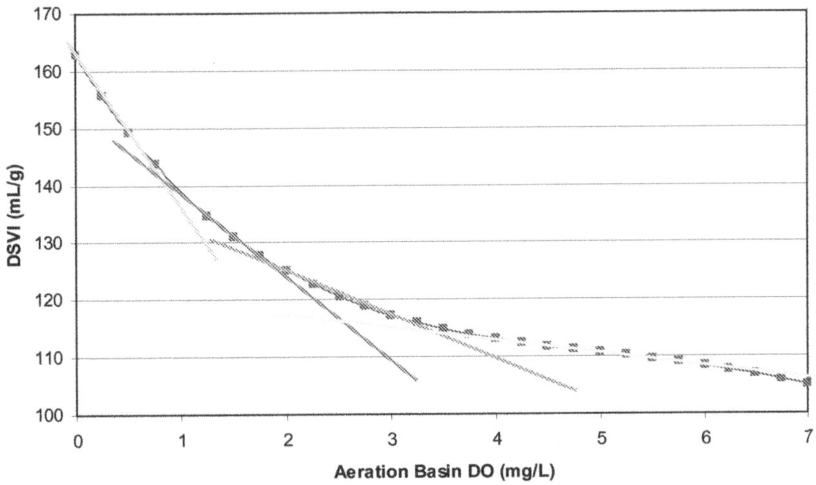

Figure 4-23b. Aeration Basin DO vs. DSVI - Cubic Polynomial Regression Curve – Short-MCRT Plants with Anoxic or Anaerobic Selectors.

Activated Sludge Influent BOD_5/TSS Ratio

According to the regression analysis, the higher the activated sludge influent BOD_5/TSS ratio, the higher the DSVI will be ($R^2 = 4.1\%$). This is consistent with sCOD, or more specifically raCOD, being a primary driver for filamentous organism proliferation in short-MCRT activated sludge systems. A selector is intended to remove the raCOD from the mixed liquor before it enters the main aeration basin, where the raCOD may be used by filamentous organisms. The higher the raCOD, the more likely the possibility that raCOD will leak through the selector and be taken up by filamentous organisms. From the cubic polynomial regression curve ($R^2 = 10.4\%$, Figures 4-24a and 4-24b), the DSVI sharply rises as BOD_5/TSS increases from 0.1 to 0.8. The DSVI rises more slowly after the BOD_5/TSS increases beyond 0.8, and then the DSVI appears to be independent of BOD_5/TSS at levels above 1.5. This illustrates the relative increase in DSVI when the BOD_5/TSS ratio increases and how the lower the BOD_5/TSS ratio is (<1.5) the better the opportunity to control filamentous bulking with an unaerated selector at short MCRTs. Since most of the data used in this regression analysis is with BOD_5/TSS ratios less than about 3.0, this relationship should not be assumed for BOD_5/TSS ratios greater than 3.0.

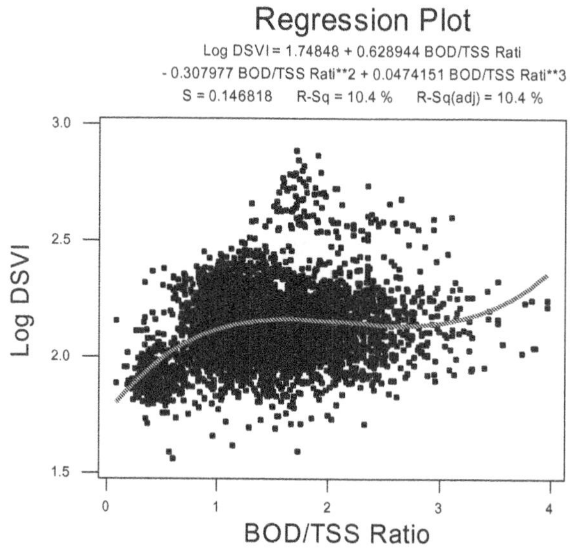

Figure 4-24a. Activated Sludge Influent BOD_5/TSS Ratio vs. Log DSVI - Cubic Polynomial Regression Plot – Short-MCRT Plants with Anoxic or Anaerobic Selectors.

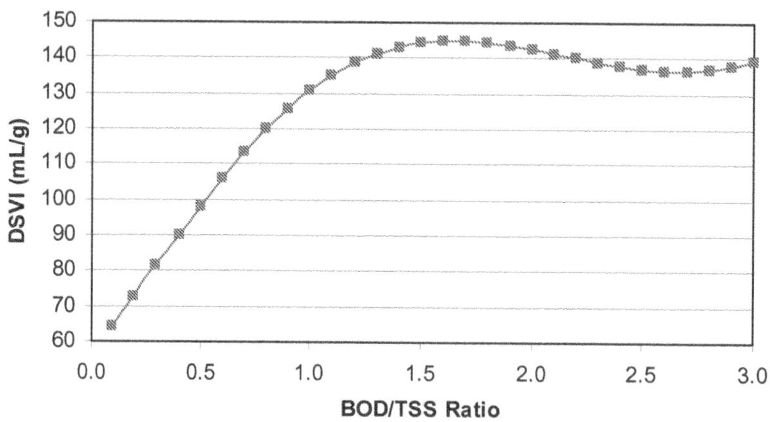

Figure 4-24b. Activated Sludge Influent BOD_5/TSS Ratio vs. DSVI - Cubic Polynomial Regression Curve – Short-MCRT Plants with Anoxic or Anaerobic Selectors.

ICZ F/M

The ICZ F/M was previously discussed under the selector F/M section. Figures 4-25a and 4-25b show that the DSVI increases sharply as the ICZ F/M increases from 0.225 to 1.5 kg BOD_5/(kg MLSS·d). Therefore the lower the ICZ F/M, the lower the DSVI. This is not consistent with the hypothesis that kinetics plays a role in the ability of an unaerated selector to control filamentous bulking. Further, using the effective ICZ volume calculated from the N value, the effective ICZ F/M linear regression R^2 value is 0.1%, and its cubic polynomial regression R^2 value is 0.9%, suggesting that the selector ICZ F/M in a short-MCRT unaerated selector does not play a major role in bulking control. Additional discussion of the ICZ F/M can be found in Appendix D.

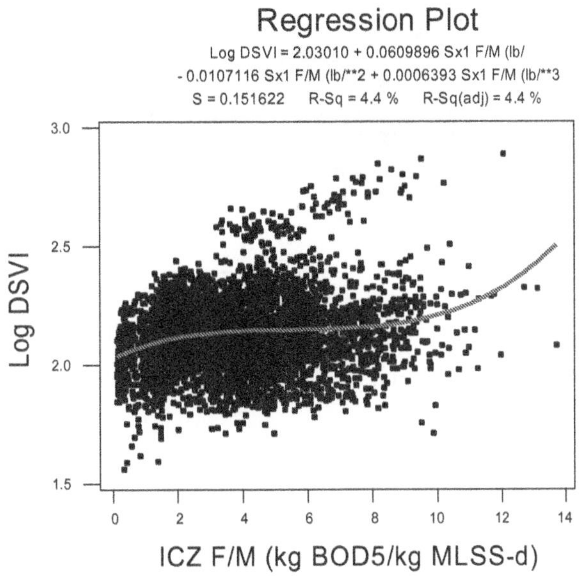

Figure 4-25a. ICZ F/M vs. Log DSVI – Cubic Polynomial Regression Plot – Short-MCRT Plants with Anoxic or Anaerobic Selectors.

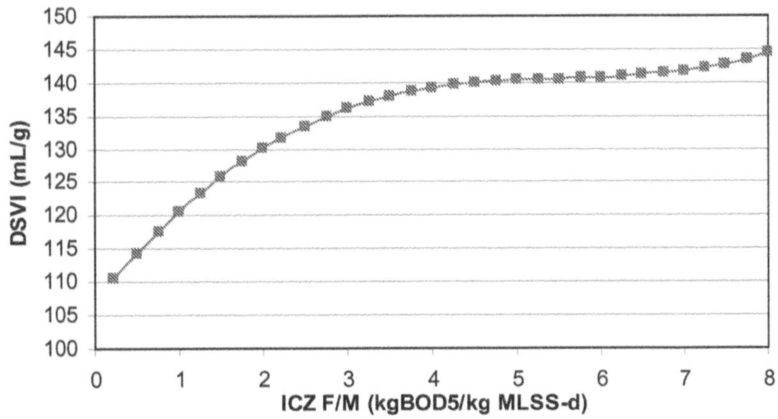

Figure 4-25b. ICZ F/M vs. DSVI – Cubic Polynomial Regression Curve – Short-MCRT Plants with Anoxic or Anaerobic Selectors.

Nominal Selector HRT (without recycle flows)

Figures 4-26a and 4-26b show that DSVI decreases as the nominal selector HRT (without mixed liquor and RAS recycle flows) increases up to 1.2 h. After 1.2 h, the DSVI appears to be constant until the nominal selector HRT is ≥2.5 h.

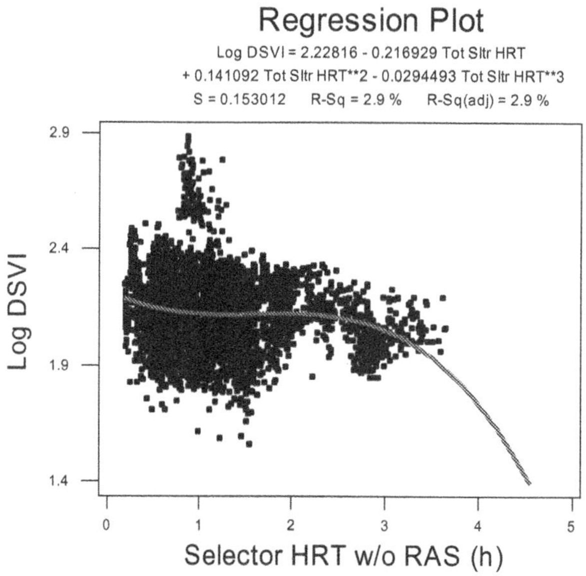

Figure 4-26a. Nominal Selector HRT (without recycle) vs. Log DSVI – Cubic Polynomial Regression Plot – Short-MCRT Plants with Anoxic or Anaerobic Selectors.

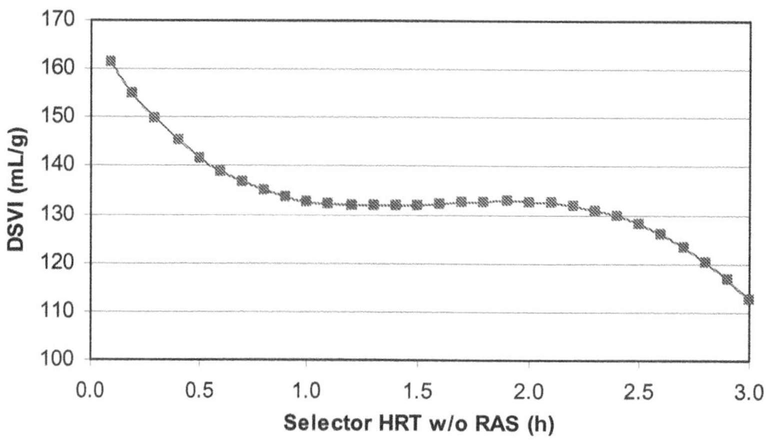

Figure 4-26b. Nominal Selector HRT (without recycle) vs. DSVI – Cubic Polynomial Regression Curve – Short-MCRT Plants with Anoxic or Anaerobic Selectors.

Selector Volume to Total Reactor Volume Ratio

According to the linear regression analysis, the DSVI decreases when the selector volume to total reactor volume ratio increases ($R^2 = 2.0\%$, t-statistic = -10.30). The quadratic regression line ($R^2 = 2.0\%$), however, shows that the DSVI decreases with increasing selector volume to reactor volume ratio until it reaches a minimum DSVI at about 0.225–0.250 (or 22.5%–25.0%, Figures 4-27a and 4-27b). The DSVI slowly increases when the selector volume to total reactor volume exceeds 0.250. This is consistent with Wanner (1994), who showed that by successively increasing the unstaged anaerobic volume to the unstaged oxic volume ratio in an anaerobic-oxic activated sludge system from 8% to 16% to 33%, the system SVI could be reduced from 800 mL/g to 250 mL/g to <100 mL/g, respectively. Converting these percentages into a selector volume to total reactor volume ratio yield: 7.4%, 13.8%, and 24.8%, which agrees very well with the polynomial regression line in Figure 4-27b.

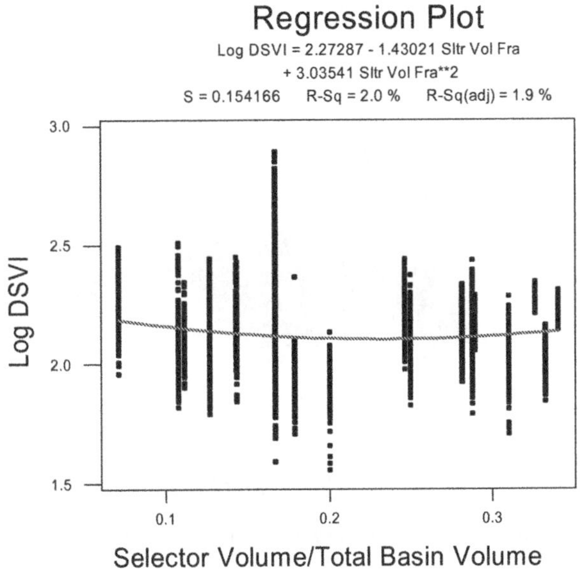

Figure 4-27a. Selector Volume to Total Basin Volume Ratio vs. Log DSVI – Cubic Polynomial Regression Plot – Short-MCRT Plants with Anoxic or Anaerobic Selectors.

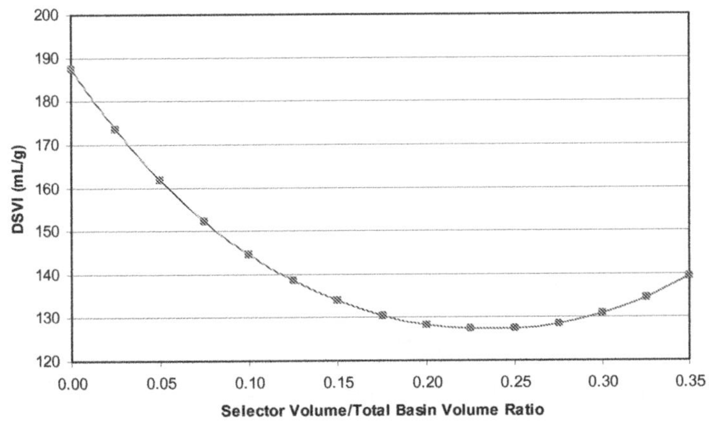

Figure 4-27b. Selector Volume to Total Basin Volume Ratio vs. DSVI – Cubic Polynomial Regression Curve – Short-MCRT Plants with Anoxic or Anaerobic Selectors.

ICZ HRT (with recycle flows)

Figures 4-28a and 4-28b show that the optimum ICZ HRT (including mixed liquor and RAS recycle flows) is approximately 1.5 h ($R^2 = 5.1\%$). The ICZ HRT (with recycle flows) was also calculated using the N value to determine the ICZ volume. The cubic regression curve was again used to determine the optimum "effective" ICZ HRT (with recycle flows, $R^2 = 5.5\%$), which was found to be >1.25 h (Figures 4-28c and 4-28d). Both ICZ HRT calculations show that the optimum ICZ HRT is approximately 1.5 h, which is about the same as that recommended for the selector HRT. This suggests that ICZ HRT is not important and that kinetics play no significant role in the ability of anoxic or anaerobic selectors to control filamentous bulking in short-MCRT plants.

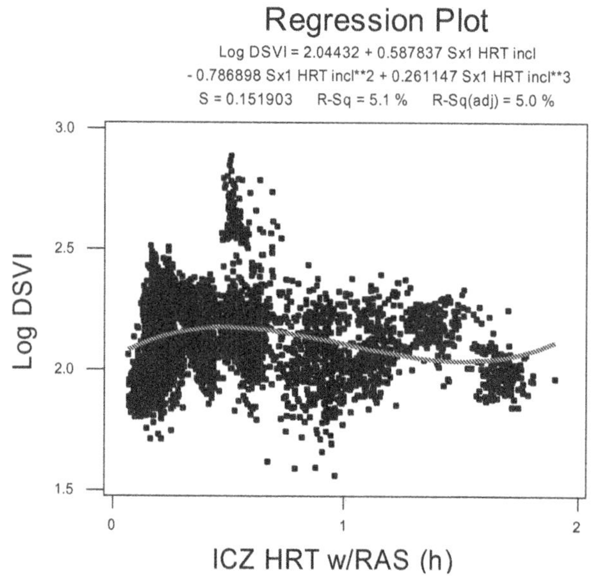

Figure 4-28a. ICZ HRT (with RAS) vs. Log DSVI – Cubic Polynomial Regression Plot – Short-MCRT Plants with Anoxic or Anaerobic Selectors.

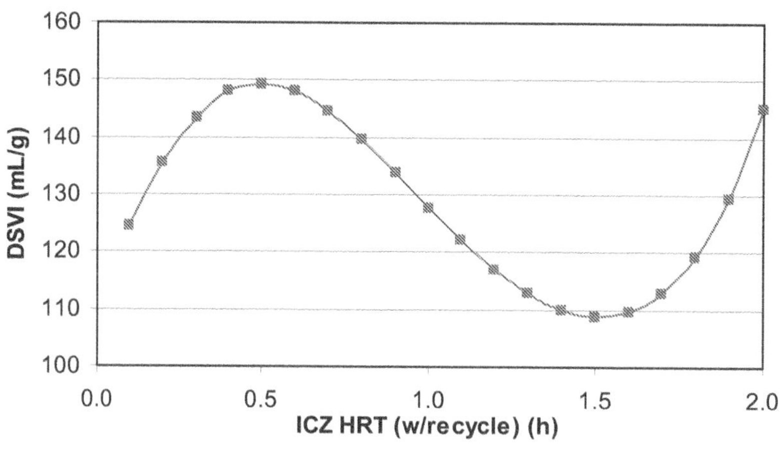

Figure 4-28b. ICZ HRT (with RAS) vs. DSVI – Cubic Polynomial Regression Curve – Short-MCRT Plants with Anoxic or Anaerobic Selectors.

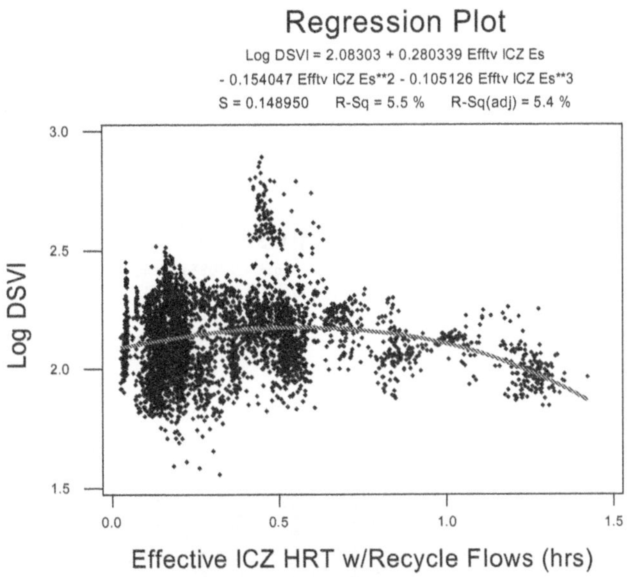

Figure 4-28c. Effective ICZ HRT (with RAS) vs. Log DSVI – Cubic Polynomial Regression Plot – Short-MCRT Plants with Anoxic or Anaerobic Selectors.

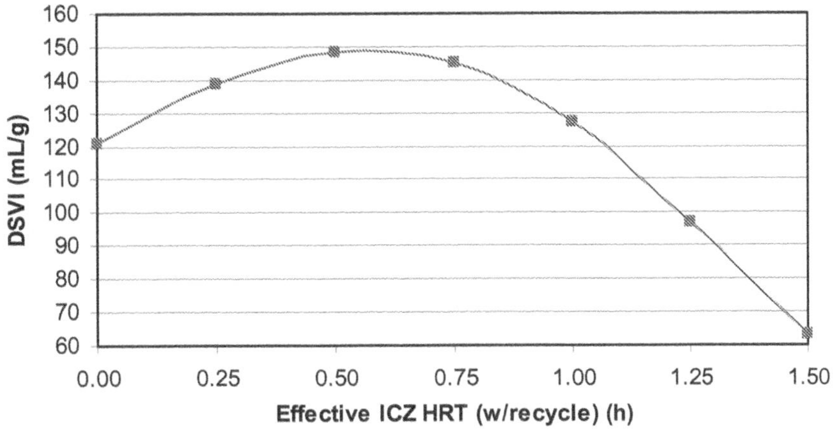

Figure 4-28d. Effective ICZ HRT (with RAS) vs. DSVI – Cubic Polynomial Regression Curve – Short-MCRT Plants with Anoxic or Anaerobic Selectors.

Effluent Temperature

Figures 4-29a and 4-29b show the cubic polynomial curve for effluent temperature regressed against DSVI (R^2=14.6%). Figure 4-29b suggests that the optimum effluent temperature for achieving low DSVIs is about 20°–25°C but that DSVI rises sharply when temperatures are 27°–30°C.

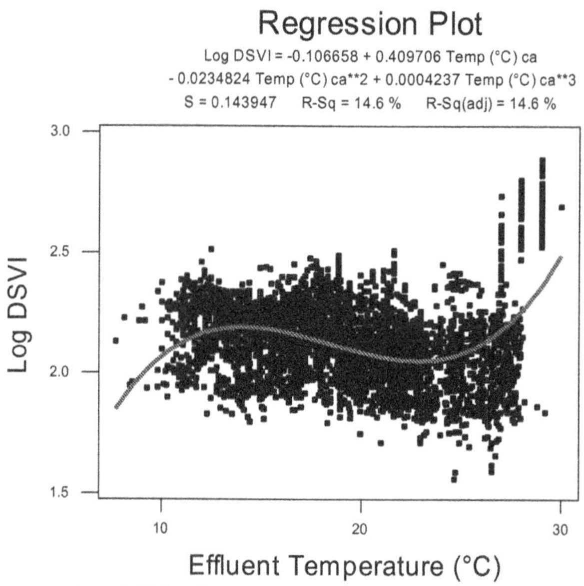

Figure 4-29a. Effluent Temperature vs. Log DSVI – Cubic Polynomial Regression Plot – Short-MCRT Plants with Anoxic or Anaerobic Selectors.

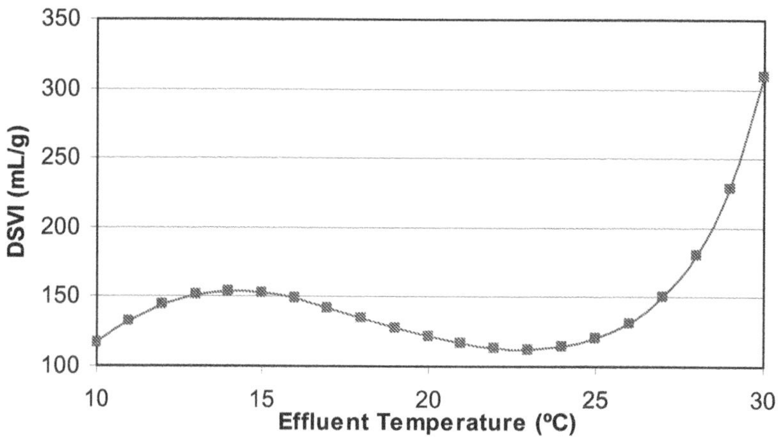

Figure 4-29b. Effluent Temperature vs. DSVI – Cubic Polynomial Regression Curve – Short-MCRT Plants with Anoxic or Anaerobic Selectors.

Insignificant Parameters

The regression analysis of the data set used in this study suggests that 1) the number of aeration basin stages, 2) the activated sludge influent BOD_5 concentration, 3) the %RAS flow, and 4) the effluent pH play no significant role in the ability of anoxic or anaerobic selectors to control filamentous bulking in short-MCRT plants.

Summary

Table 4-15 summarizes the best ranges found in this study (as demonstrated by the regression analysis) and the literature for the most important design and operating parameters for controlling filamentous bulking in short-MCRT anoxic or anaerobic selector activated sludge systems.

Table 4-15. Recommended Parameter Ranges for Short-MCRT Plants with Anoxic or Anaerobic Selectors.

Parameter	Recommendations from this Study	Recommendations from Literature	Literature References
Average MLSS (mg/L)	1,500-2,000+		
Reactor MCRT (d)	>4.5		
Total Selector F/M [kg BOD_5/(kg MLSS·d)]	<1.0 (lower is better)	≤1.0	Jenkins, 2004
Selector MCRT (d)	2-3+	1.0-2.0	Jenkins, 2004
Number of Selector Stages	2	3	Jenkins, 2004; Wanner, 1994
Aeration Basin DO (mg/L)	2.5-4.0 (air plants only)	>1-2	Jenkins, 2004; Wanner, 1994
Activ. Sludge Influent BOD/TSS Ratio[1]	<0.5 (lower is better)		
ICZ F/M [kg BOD_5/(kg TSS·d)]	<1.0 (lower is better)	6	Wanner, 1994
Selector HRT (without recycle) (h)	min. of 1.2, >2.5 best		
Selector Vol/Total Basin Vol Ratio (%)	22.5-25.0	25	Wanner, 1994
Selector HRT (with recycle) (h)	>1.5	0.75-2.0	Jenkins, 2004
ICZ HRT with RAS (h)	1.4-1.6		
ICZ HRT without RAS (h)	2.4-2.7		
Effluent Temperature (°C) [1]	20-25 (27-30+ worst)		
Number of Aeration Basin Stages	not significant		
Act Sldg. Inf. BOD (mg/L)	not significant		
%RAS Flow (%)	not significant	≤100	Wanner, 1994
Effluent pH	not significant		

Note:
[1] Best results found in this range, but not recommended to make adjustments to operate in this range.

4.4.11.2 Short-MCRT Plants with Aerobic Selectors: Regression Results

Activated Sludge Influent BOD_5 Concentration

The activated sludge influent BOD_5 concentration had the greatest influence on DSVI in the short-MCRT plants with aerated selectors, as indicated by the linear regression $R^2 = 36.7\%$. Figures 4-30a and 4-30b show the cubic polynomial regression curve. From Figure 30b, the lowest DSVIs occurred at influent BOD_5 values <80 mg/L; influent BOD_5 values of 80–100 mg/L were the next best; influent BOD_5 values at 120–140 mg/L started to yield higher DSVIs; and influent BOD_5 values >150 mg/L resulted in the highest DSVIs.

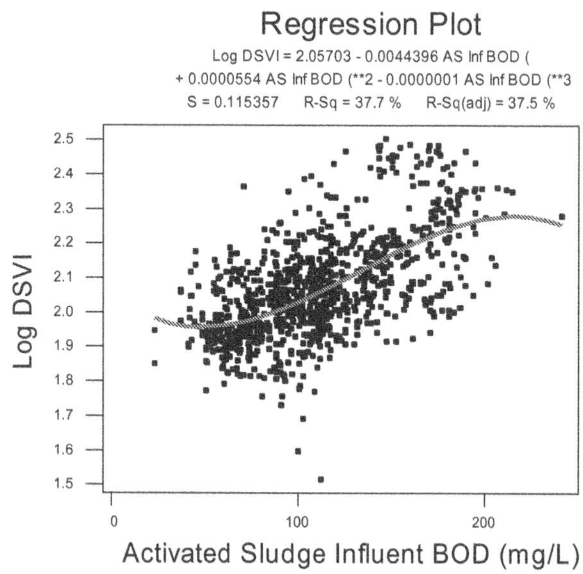

Figure 4-30a. Activated Sludge Influent BOD vs. Log DSVI – Cubic Polynomial Regression Plot – Short-MCRT Plants with Aerobic Selectors.

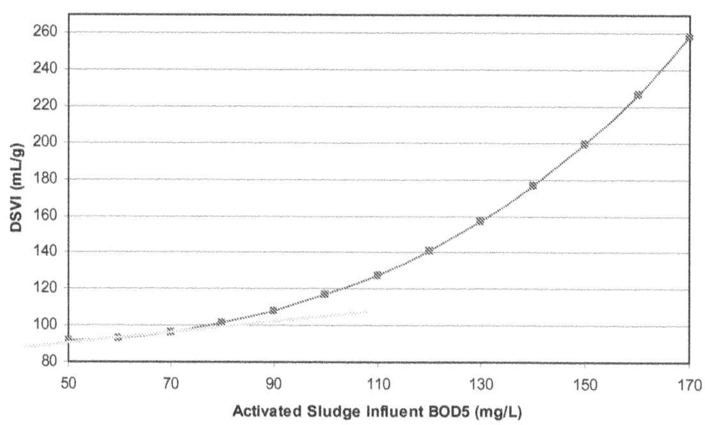

Figure 4-30b. Activated Sludge Influent BOD vs. DSVI – Cubic Polynomial Regression Curve – Short-MCRT Plants with Aerobic Selectors.

Nominal ICZ HRT (without recycle flows)

The nominal ICZ HRT (without mixed liquor and RAS recycle flows) was almost as important in its influence on DSVI (R^2 = 33.7%) as the influent BOD_5 concentration. DSVI increases with rising ICZ HRT. The polynomial regression curve (Figures 4-31a and 4-31b) shows that the optimum ICZ HRT is about 0.09–0.11 h or 5.4–6.6 min. A greater operating range of 4.5–7.5 min could be used for better practicality. This demonstrates that ICZs are important to short-MCRT aerated selectors and that kinetics play a major role. The effective ICZ HRT (calculated using the N value) without recycle flows also had strong influence on DSVI, with the linear R^2 = 28.3% and t-statistic = 20.1, further supporting that kinetics play a major role in the ability of an aerobic selector to control bulking in short-MCRT plants.

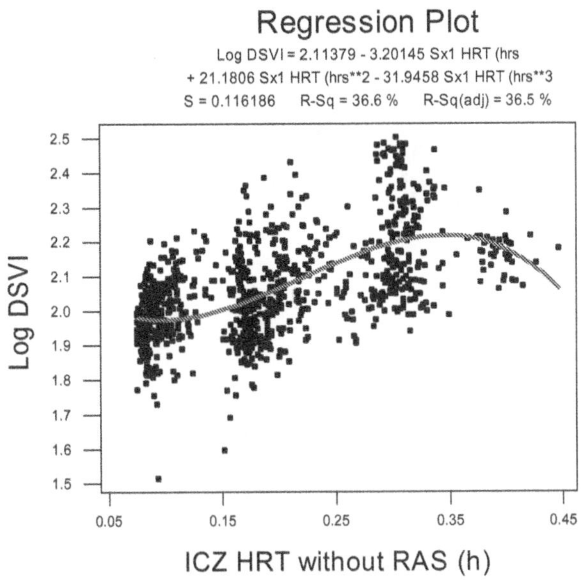

Figure 4-31a. Nominal ICZ HRT (without recycle) vs. Log DSVI – Cubic Polynomial Regression Plot – Short-MCRT Plants with Aerobic Selectors.

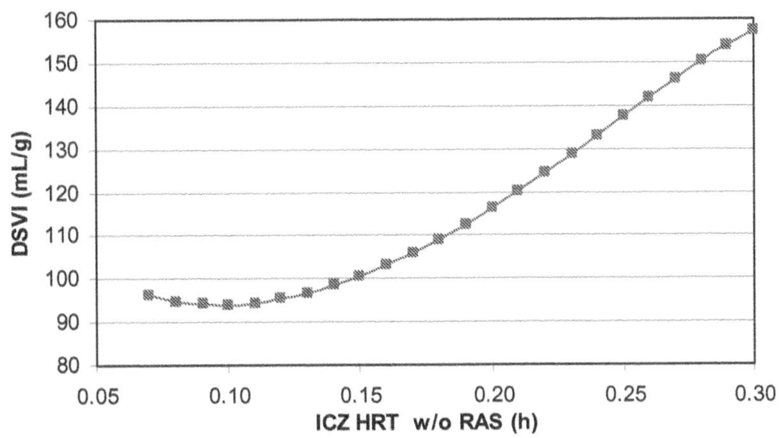

Figure 4-31b. Nominal ICZ HRT (without recycle) vs. DSVI – Cubic Polynomial Regression Curve – Short-MCRT Plants with Aerobic Selectors.

ICZ HRT (with recycle flows)

The ICZ HRT (including mixed liquor and RAS recycle flows) is the next most influential parameter on DSVI with a linear regression $R^2 = 30.8\%$. Figures 4-32a and 4-32b show that the optimum ICZ HRT (including recycle flows) is 0.06–0.10 h or about 3.5–6.0 min. The effective ICZ HRT (including recycle flows) is also very significant with a linear regression $R^2 = 23.5\%$.

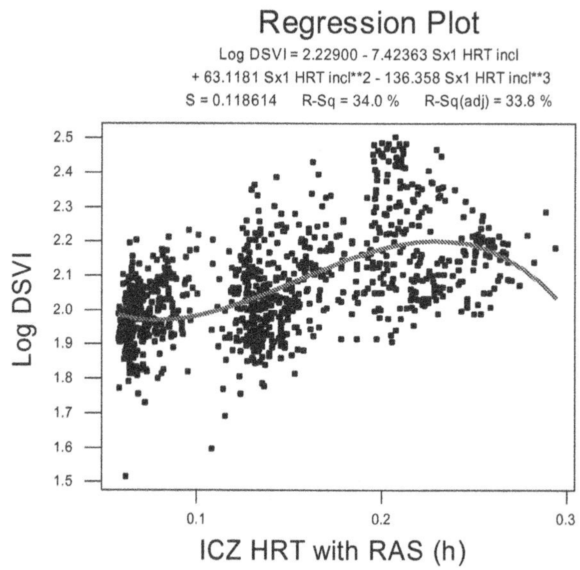

Figure 4-32a. ICZ HRT (with recycle) vs. Log DSVI – Cubic Polynomial Regression Plot – Short-MCRT Plants with Aerobic Selectors.

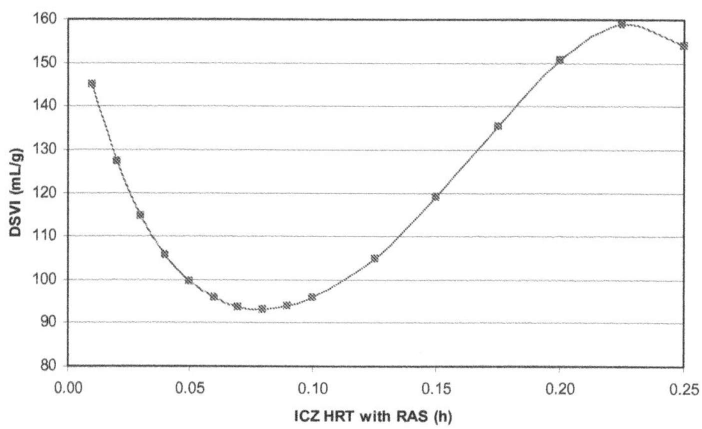

Figure 4-32b. ICZ HRT (with recycle) vs. DSVI – Cubic Polynomial Regression Curve – Short-MCRT Plants with Aerobic Selectors.

Effluent pH

Effluent pH had a significant influence on DSVI with a linear regression $R^2 = 29.4\%$. Figures 4-33a and 4-33b show the cubic polynomial regression for this parameter regressed against DSVI ($R^2 = 35.4\%$). Figure 33b shows the DSVI is lowest at pH = 6.3–6.6, and the highest DSVI occurred at pH >7.3. Except for bulking caused by filamentous fungi, the literature does not discuss the effects of pH on filamentous bulking. However, Pellegrin et al. (1999) report that *S. natans* exhibited reduced growth at pH = 5.4–6.3, and Howarth et al. (1999) showed that *Thiothrix* grew between a pH range of 6.5 to 8.5. It is assumed that reduced or no *Thiothrix* growth occurred outside of this pH range.

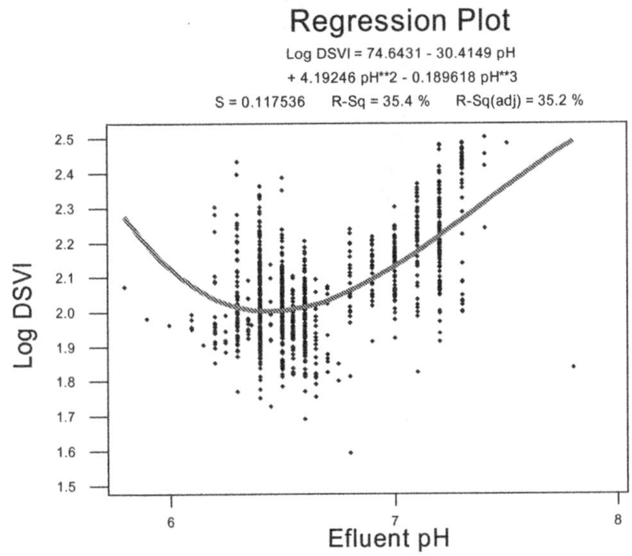

Figure 4-33a. Effluent pH vs. Log DSVI – Cubic Polynomial Regression Plot – Short-MCRT Plants with Aerobic Selectors.

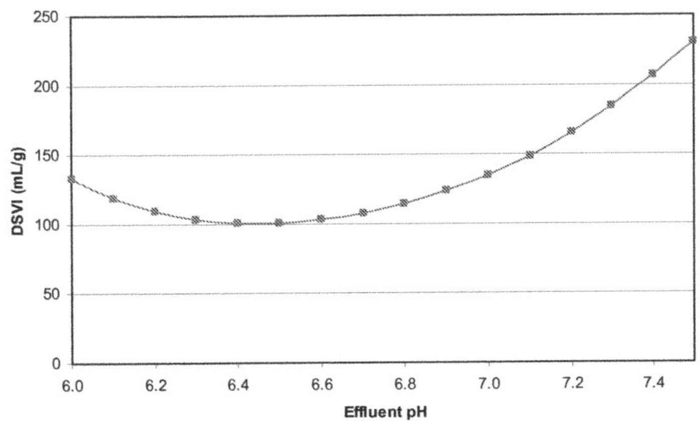

Figure 4-33b. Effluent pH vs. DSVI – Cubic Polynomial Regression Curve – Short-MCRT Plants with Aerobic Selectors.

Total Selector HRT (without recycle flows)

Figures 4-34a and 4-34b show the cubic polynomial regression curve of the total selector HRT (without mixed liquor and RAS recycle flows) plotted against DSVI ($R^2 = 31.2\%$). Figure 32b shows that the lowest DSVI in the group of short-MCRT plants with aerobic selectors occurred at a selector HRT = 0.30 h or 18 min.

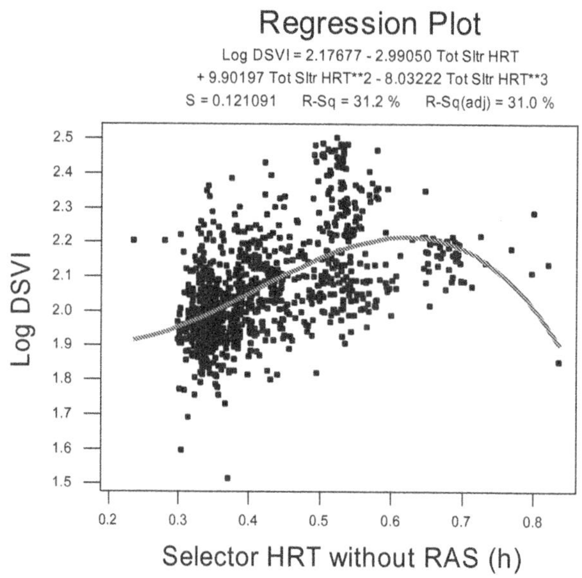

Figure 4-34a. Nominal Selector HRT (without recycle) vs. Log DSVI – Cubic Polynomial Regression Plot – Short-MCRT Plants with Aerobic Selectors.

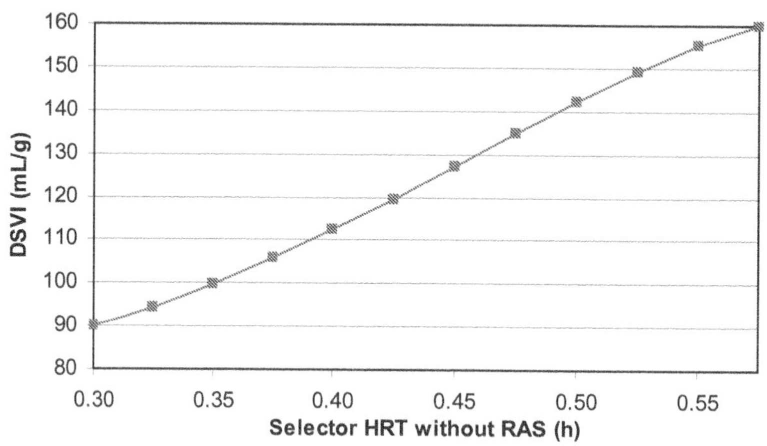

Figure 4-34b. Nominal Selector HRT (without recycle) vs. DSVI – Cubic Polynomial Regression Curve – Short-MCRT Plants with Aerobic Selectors.

Percent RAS Flow Returned to the Selector

Figures 4-35a and 4-35b show the cubic polynomial regression curve for the %RAS flow regressed against the DSVI (R^2 = 24.9%). Figure 4-35b shows the optimum %RAS flow is 25%–35% for the short-MCRT aerated selector plants. When the actual plant data was examined, it was discovered that the %RAS flow was increased prior to the DSVI rise, showing that the higher %RAS was responsible in part for the DSVI rise in some cases. This may be due to a possible dilution effect caused by the higher RAS flow.

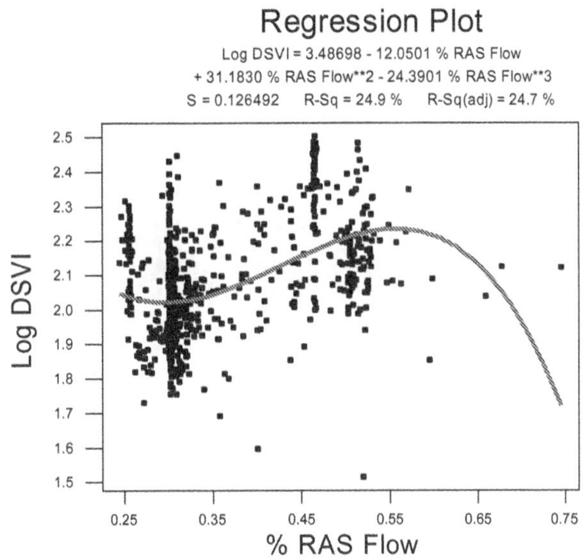

Figure 4-35a. Percent RAS Flow vs. Log DSVI – Cubic Polynomial Regression Plot – Short-MCRT Plants with Aerobic Selectors.

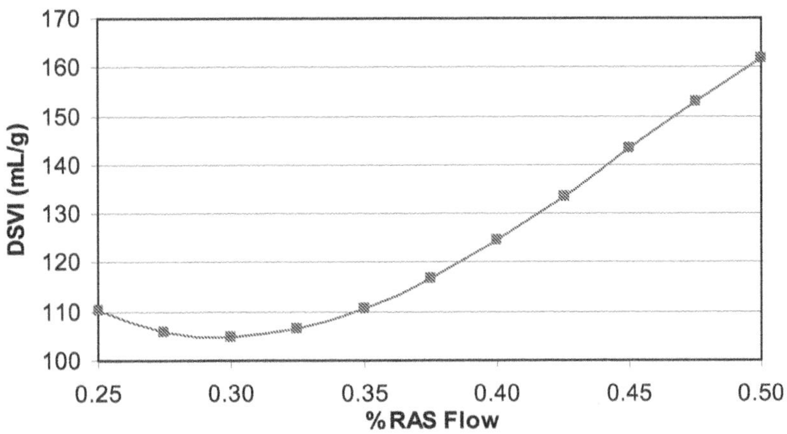

Figure 4-35b. Percent RAS Flow vs. DSVI – Cubic Polynomial Regression Curve – Short-MCRT Plants with Aerobic Selectors.

Effluent Temperature

Figures 4-36a and 4-36b show the cubic polynomial regression curve for the effluent temperature regressed against DSVI (R^2 = 18.4%). Figure 4-36b shows that the lower the temperature the better, down to about 12°C for the short-MCRT plants with aerobic selectors. The DSVI, however, appears to increase at a faster rate above about 18°–19°C, but then the DSVI sharply rises between about 21°-23°C.

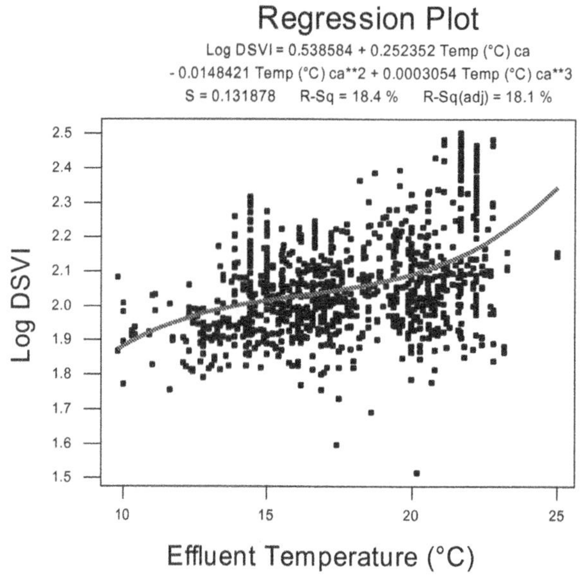

Figure 4-36a. Effluent Temperature vs. Log DSVI – Cubic Polynomial Regression Plot – Short-MCRT Plants with Aerobic Selectors.

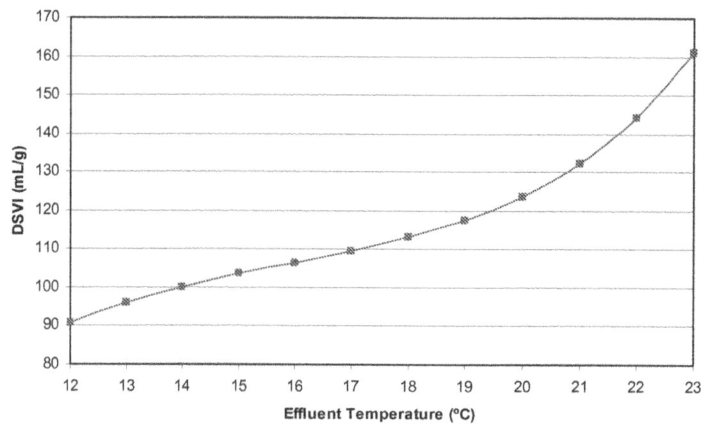

Figure 4-36b. Effluent Temperature vs. DSVI – Cubic Polynomial Regression Curve – Short-MCRT Plants with Aerobic Selectors.

ICZ F/M

Figures 4-37a and 4-37b show the cubic polynomial regression curve for the ICZ F/M regressed against DSVI. Figure 4-37b shows that DSVI drops rapidly at an ICZ F/M of 12–20 kg BOD_5/(kg MLSS·d), which further demonstrates the importance of kinetics on the success of a short-MCRT aerated selector to control filamentous bulking. The effective ICZ F/M (calculated with the N value) had a much lower linear regression R^2 value (1.6%), which more clearly demonstrates that the ICZ HRT is a more important design and operating parameter than ICZ F/M in aerated short-MCRT selectors (recalling that the linear regression $R^2 = 28.3\%$ for the effective ICZ HRT).

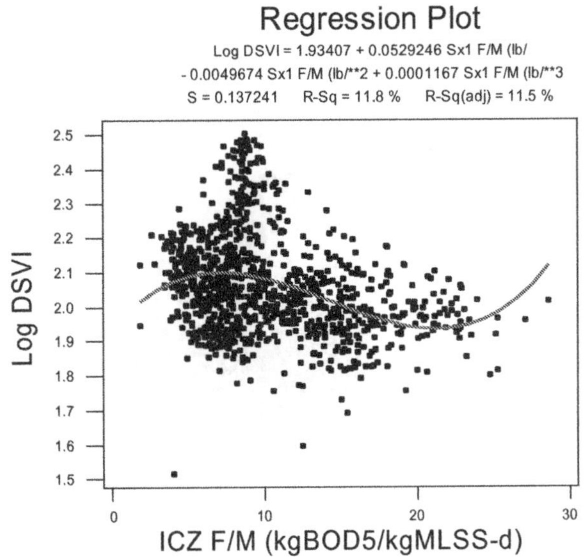

Figure 4-37a. ICZ F/M vs. Log DSVI – Cubic Polynomial Regression Plot – Short-MCRT Plants with Aerobic Selectors.

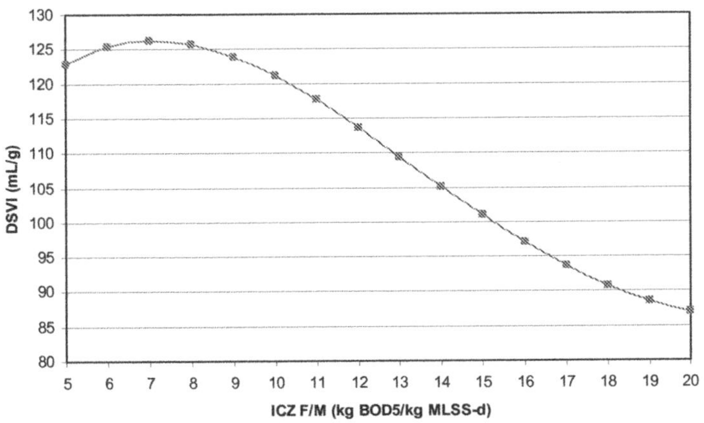

Figure 4-37b. ICZ F/M vs. DSVI – Cubic Polynomial Regression Curve – Short-MCRT Plants with Aerobic Selectors.

MLSS

Contrary to the short-MCRT plants with anoxic or anaerobic selectors (and the long-MCRT plants), MLSS is not the most important parameter for short-MCRT aerated selectors (linear regression $R^2 = 4.7\%$). Figures 4-38a and 4-38b show the cubic polynomial regression curve for the MLSS regressed against the DSVI. The lower the MLSS drops, the lower the DSVI (also contrary to the other plant groups). Since the lowest MLSS value collected for plants in this group was about 1,000 mg/L, this value is recommended to achieve lower DSVIs in short-MCRT plants with aerobic selectors. The lower MLSS might offer an advantage because it imparts a lower oxygen demand in the aerated selector, where the oxygen uptake rates are very high and dissolved oxygen is at a premium.

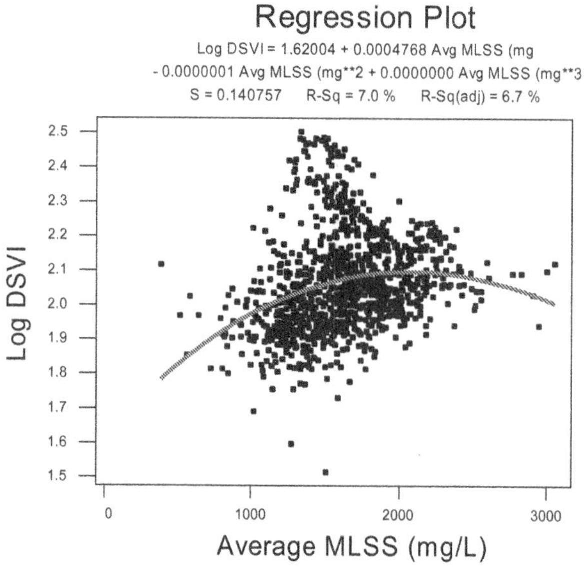

Figure 4-38a. MLSS vs. Log DSVI – Cubic Polynomial Regression Plot – Short-MCRT Plants with Aerobic Selectors.

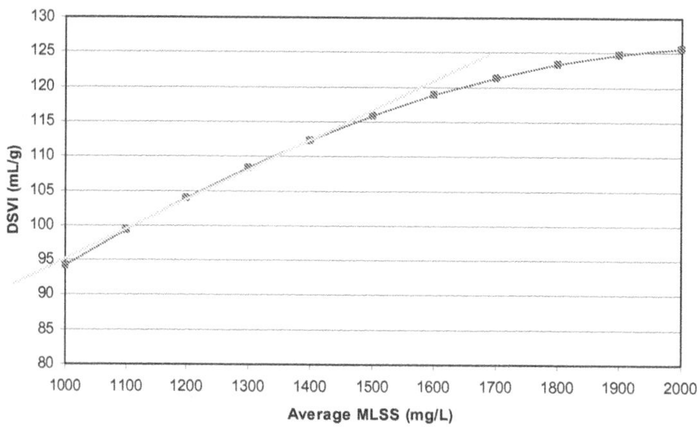

Figure 4-38b. MLSS vs. DSVI – Cubic Polynomial Regression Curve – Short-MCRT Plants with Aerobic Selectors.

Summary

Table 4-16 summarizes the best ranges found in this study (as shown by the regression analysis) and the literature for the most important design and operating parameters for controlling filamentous bulking in short-MCRT activated sludge systems with aerobic selectors.

Table 4-16. Recommended Parameter Ranges for Short-MCRT Plants with Aerobic Selectors.

Parameter	Recommendations from this Study	Recommendations from Literature	Literature References
Act. Sldg. Inf. BOD (mg/L)	<80	N/A	
ICZ HRT (without recycle) (min)	4.5-7.5		
ICZ HRT (with recycle) (min)	3.5-6.0		
Effluent pH	6.3-6.6	N/A	
Total Selector HRT (without recycle) (min)	≤18		
% RAS Flow (%)	25-35	≤100	Wanner, 1994
Total Selector HRT with RAS (min)	15-18	10-20	Wanner, 1994
Effluent Temperature (°C)	<18-19 (worst: 21-23+)	<28	Wanner, 1994
ICZ F/M [kg BOD_5/(kg MLSS·d)]	~15	~5-6 ≥16 ok	Jenkins et al., 2004 Wanner, 1994
Reactor MCRT (d)	<1.3		
Average MLSS (mg/L)	max. of 1,000		
Aeration Basin DO (mg/L)	14-18 (pure O_2 plants)	>10 (pure O_2 plants)	Wanner, 1994
Total Selector F/M [kg BOD_5/(kg MLSS·d)]	not significant	~1.5-2.0	Jenkins et al., 2004
Number of Selector Compartments	N/A[1]	3	Wanner, 1994; Jenkins, 2004

Note:
[1] Insufficient data variation in data set to adequately assess this parameter.

4.4.11.3 Long-MCRT Plants with Selectors: Regression Results

Average MLSS

The average MLSS shows the highest correlation to log DSVI, with the highest R^2 value at 23.4% (t-statistic = -52.69) compared to all other parameters tested in this study for the long-MCRT plants. Figures 4-39a and 4-39b show the cubic polynomial regression curve (R^2 = 25.0%) for MLSS regressed against DSVI. Figure 4-39b shows the steepest DSVI drop between 500 and about 2,500 mg/L. The DSVI continues to drop at a lower rate between 2,500 and 4,500 mg/L, and then the curve starts to flatten beyond 4,500 mg/L.

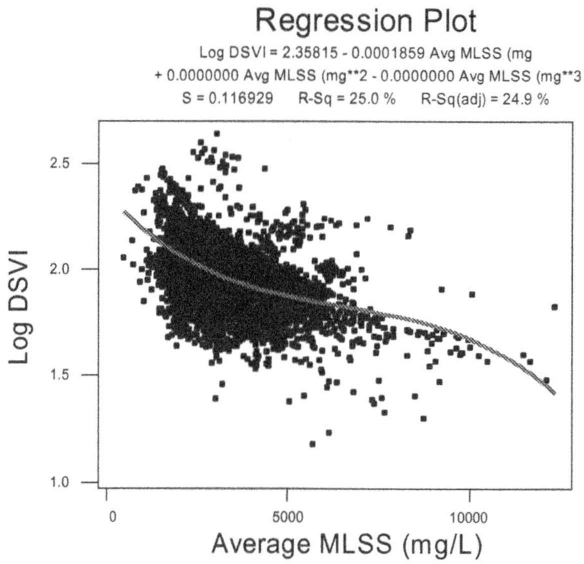

Figure 4-39a. MLSS vs. Log DSVI – Cubic Polynomial Regression Plot – Long-MCRT Plants with Selectors.

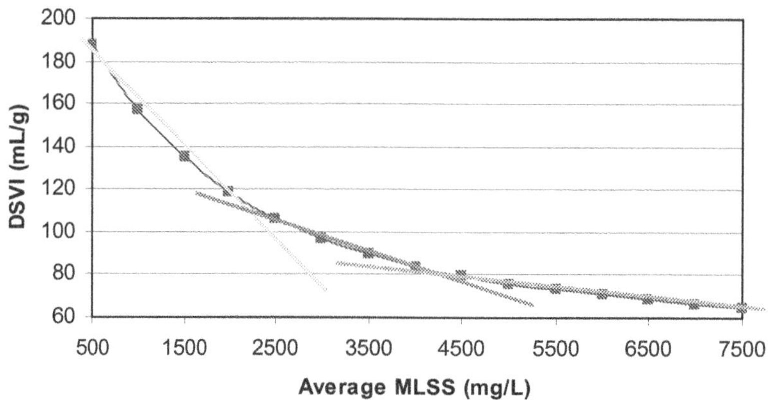

Figure 4-39b. MLSS vs. DSVI – Cubic Polynomial Regression Curve – Long-MCRT Plants with Selectors.

Selector HRT (with recycle flows)

The selector HRT (including mixed liquor and RAS recycle flows) had the second highest influence on DSVI for the long-MCRT plants with a linear regression $R^2 = 11.5\%$ and t-statistic = 30.55. This means that the lower the selector HRT the lower the DSVI, which is similar to the short-MCRT plants with aerobic selectors. Figures 4-40a and 4-40b, however, show that the cubic polynomial regression curve is significantly different from that for the short-MCRT plants with aerobic selectors. Figure 4-40b shows that the DSVI is lowest when the selector HRT approaches zero hours (→ 0 h), which suggests that no selector is better than a selector of any HRT. Nearly all (23 of 24) of the long-MCRT plants used in this study group had anoxic selectors, so this analysis may suggest that filamentous bulking can be reduced by eliminating anoxic zones in long-MCRT activated sludge systems.

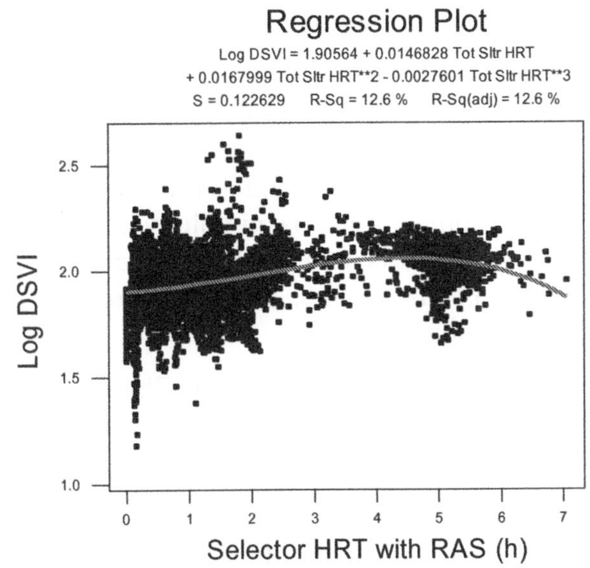

Figure 4-40a. Selector HRT (with recycle) vs. Log DSVI – Cubic Polynomial Regression Plot – Long-MCRT Plants with Selectors.

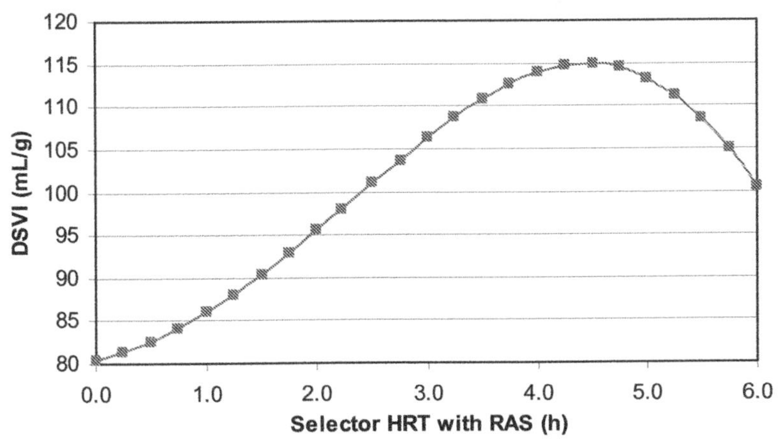

Figure 4-40b. Selector HRT (with recycle) vs. DSVI – Cubic Polynomial Regression Curve – Long-MCRT Plants with Selectors.

ICZ HRT (with recycle flows)

The ICZ HRT (including mixed liquor and RAS recycle flows) is the next parameter most correlated with DSVI ($R^2 = 10.8\%$) for the long-MCRT plants, and the regression analysis results also show, like the selector HRT, that low DSVIs occur at an ICZ HRT → 0 h. The effective ICZ HRT with recycle flows included (calculated with the N value) regressed against log DSVI results in a linear regression $R^2 = 3.2\%$. The effective ICZ HRT may be less tied to the selector HRT and therefore may be more accurate. Nonetheless, in either case, the lowest DSVI occurs when the ICZ HRT → 0 h. Since the $R^2 = 0.1\%$ for ICZ F/M, this parameter is not significant. The effective ICZ F/M has a slightly higher influence on DSVI, but the R^2 is still low at 0.4%. These results suggest that kinetics play little, if any, role in the success of selectors in long-MCRT activated sludge systems.

Selector Volume to Total Reactor Volume Ratio

According to the linear regression analysis, the DSVI increases when the selector volume to total reactor volume ratio increases ($R^2 = 5.9\%$, t-statistic = 21.36) (refer to Table 4-14) in long-MCRT plants. This is in contrast to the regression analysis results for short-MCRT plants with anoxic or anaerobic selectors. Figures 4-41a and 4-41b show the cubic polynomial regression curve for the selector volume to total basin volume ratio regressed against DSVI ($R^2 = 6.4\%$). Figure 4-41b shows that the DSVI is lowest when the selector volume to total basin volume ratio → 0 (i.e., when there is no selector). This corresponds well with the selector HRT regression results discussed previously.

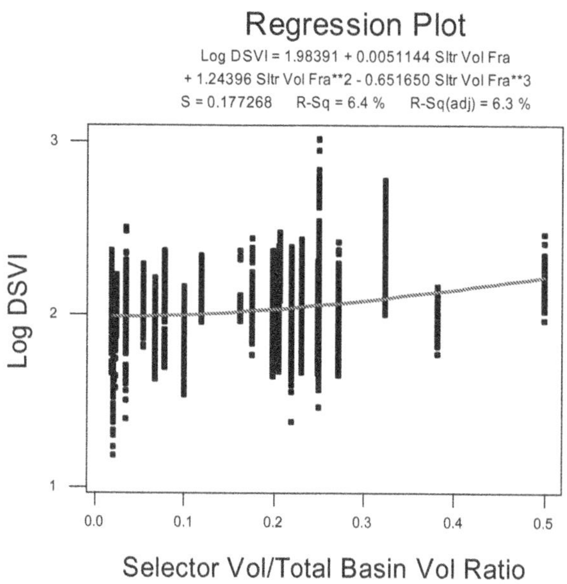

Figure 4-41a. Selector Volume to Total Basin Volume Ratio vs. Log DSVI – Cubic Polynomial Regression Plot – Long-MCRT Plants with Selectors.

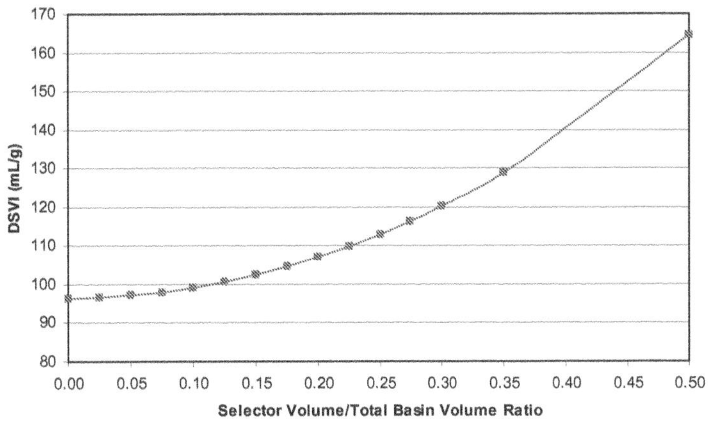

Figure 4-41b. Selector Volume to Total Basin Volume Ratio vs. DSVI – Cubic Polynomial Regression Curve – Long-MCRT Plants with Selectors.

Number of Aeration Basin Stages

Unlike the short-MCRT plants with anoxic or anaerobic selectors group, the higher the number of aeration basin stages the lower the DSVI for the long-MCRT group. Figures 4-42a and 4-42b show the cubic polynomial regression curve for the number of aeration basin stages regressed against DSVI ($R^2 = 4.9\%$). Figure 42b shows that the polynomial curve is essentially a straight line showing the lowest DSVI when there are eight aeration basin stages (highest in this study group). The additional aeration basin stages may provide for better aeration and improved control of filamentous organisms and bulking.

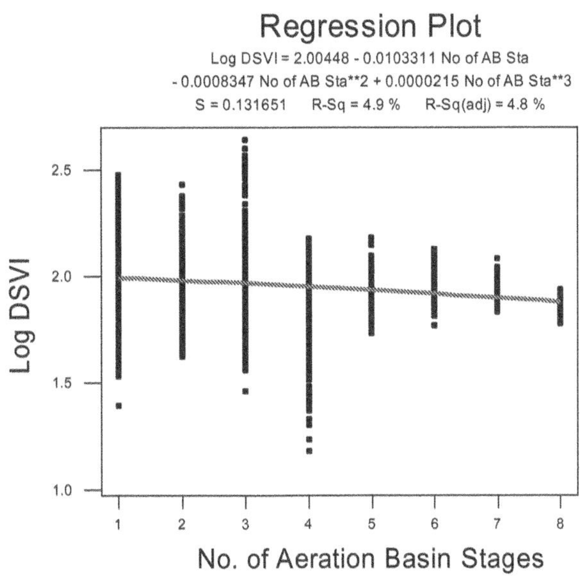

Figure 4-42a. Number of Aeration Basin Stages vs. Log DSVI – Cubic Polynomial Regression Plot – Long-MCRT Plants with Selectors.

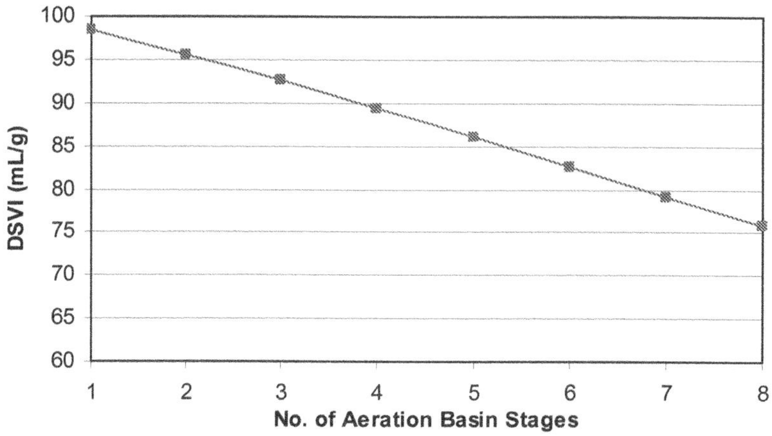

Figure 4-42b. Number of Aeration Basin Stages vs. DSVI – Cubic Polynomial Regression Curve – Long-MCRT Plants with Selectors.

Number of Selector Stages

In contrast to the number of aeration basin stages, the lower the number of selector stages the lower the DSVI for long-MCRT plants. Figures 4-43a and 4-43b show the cubic polynomial regression curve for the number of selector stages regressed against the DSVI ($R^2 = 4.2\%$). Figure 4-43b shows that the lowest DSVI is when the number of selector stages = 0. This is similar to the ICZ and selector HRTs. Figures 4-43c and 4-43d show the cubic polynomial regression curve for the effective number of selector stages (calculated using the N value). The cubic regression R^2 is actually higher for the effective number of selector stages ($R^2 = 6.4\%$) compared to that of the number of selector stages. Nonetheless, Figure 4-43d also shows the lowest DSVI occurs when the number of selector stages = 0.

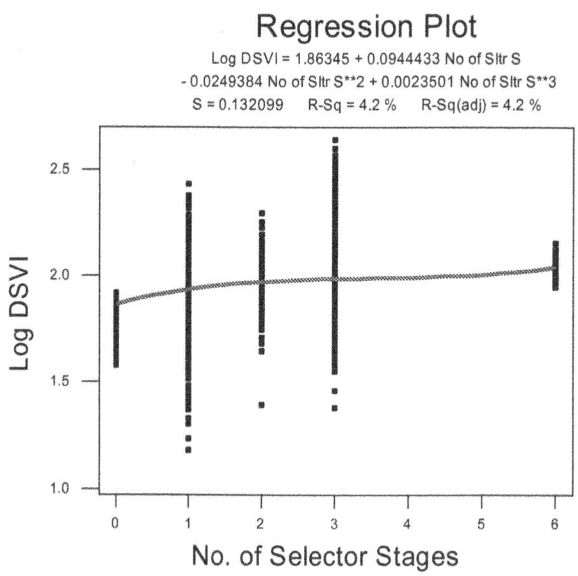

Figure 4-43a. Number of Selector Stages vs. Log DSVI – Cubic Polynomial Regression Plot – Long-MCRT Plants with Selectors.

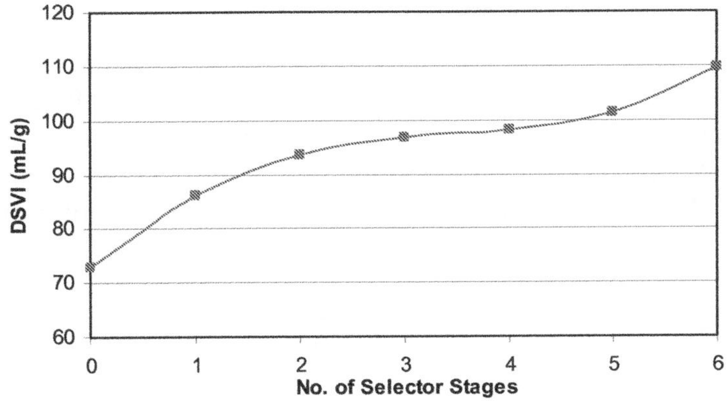

Figure 4-43b. Number of Selector Stages vs. DSVI – Cubic Polynomial Regression Curve – Long-MCRT Plants with Selectors.

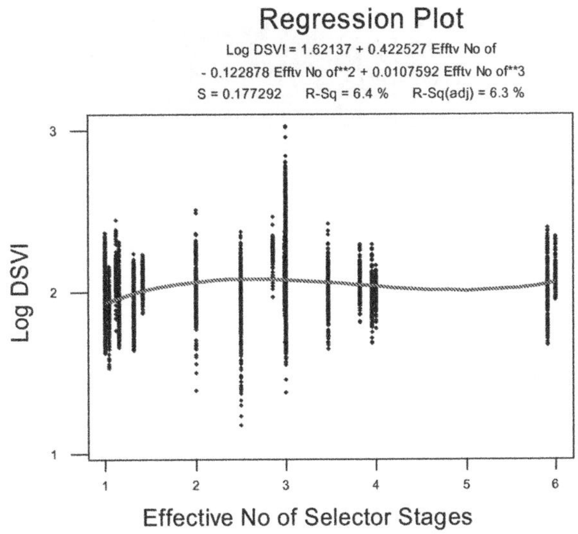

Figure 4-43c. Number of Effective Selector Stages vs. Log DSVI – Cubic Polynomial Regression Plot – Long-MCRT Plants with Selectors.

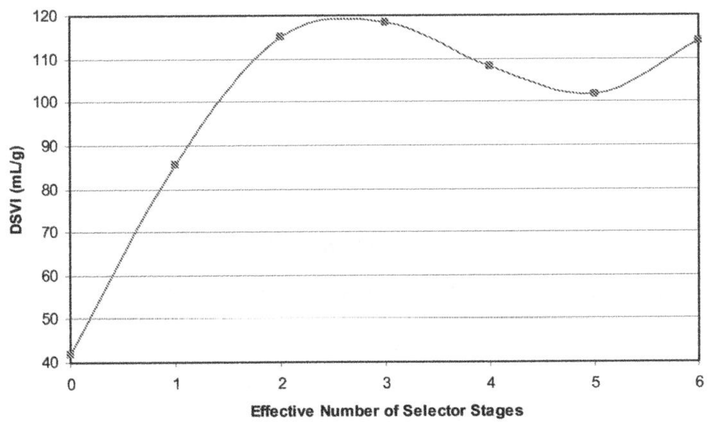

Figure 4-43d. Number of Effective Selector Stages vs. DSVI – Cubic Polynomial Regression Curve – Long-MCRT Plants with Selectors.

Effluent pH

Figures 4-44a and 4-44b show the cubic polynomial regression curve for effluent pH regressed against DSVI. Figure 4-44b shows that the DSVI for long-MCRT plants is lowest when the pH is 6.4–6.7, and the DSVI is highest when the pH is about 7.7–8.0. This appears to correspond well with Wanner (1994) who reported that *M. parvicella*—a filamentous bacteria that grows best at long MCRTs—grew best at pH = 7.7–8.0, but did not grow at pH <7.1.

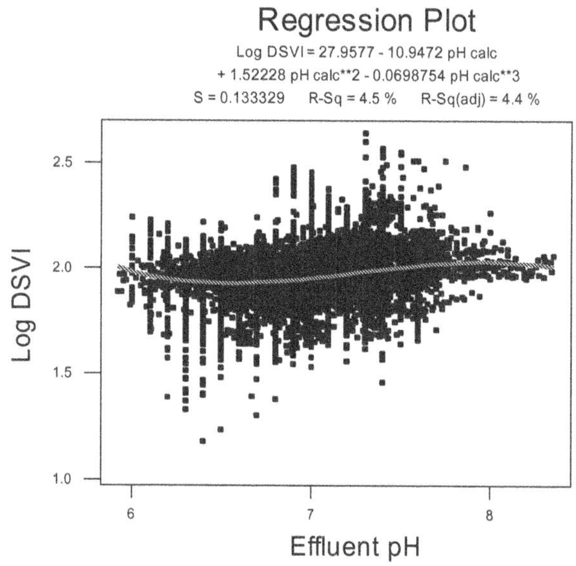

Figure 4-44a. Effluent pH vs. Log DSVI – Cubic Polynomial Regression Plot – Long-MCRT Plants with Selectors.

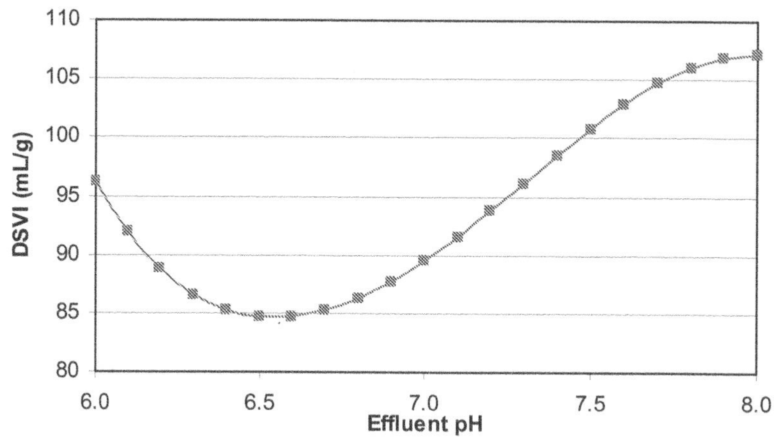

Figure 4-44b. Effluent pH vs. DSVI – Cubic Polynomial Regression Curve – Long-MCRT Plants with Selectors.

Basin Effluent Temperature

Basin effluent temperatures ranged from about 5°C to about 35°C for data used in this regression analysis. Within this temperature range, DSVI dropped when temperature increased (R^2 = 2.8%, t-statistic = -15.56) (Table 4-14). Figures 4-45a and 4-45b show the cubic polynomial curve for basin effluent temperature regressed against DSVI (R^2 = 4.7%). Figure 4-45b shows that DSVI is highest between 13°–17°C, and lowest between 27°–32°C (also lowest between 5°–7°C). In general, long-MCRT filamentous bacteria have not been studied as much as short-MCRT filamentous bacteria with the exception of *M. parvicella*. *M. parvicella* has been reported to grow best at temperatures less than 12°–15°C, and not at all at 35°C (Wanner, 1994; Jenkins et al., 2004).

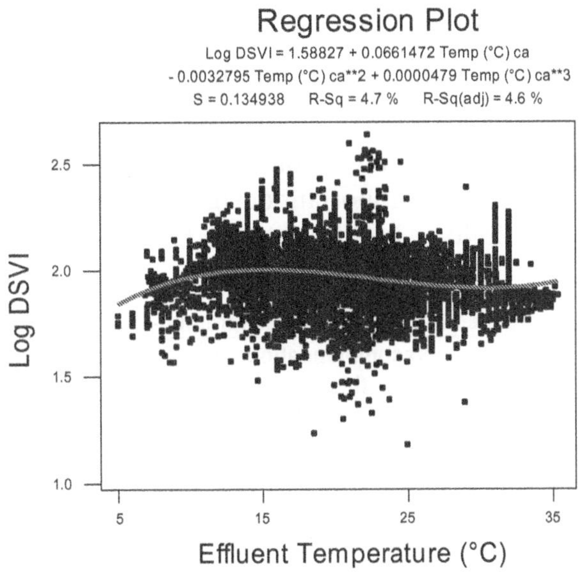

Figure 4-45a. Effluent Temperature vs. Log DSVI – Cubic Polynomial Regression Plot – Long-MCRT Plants with Selectors.

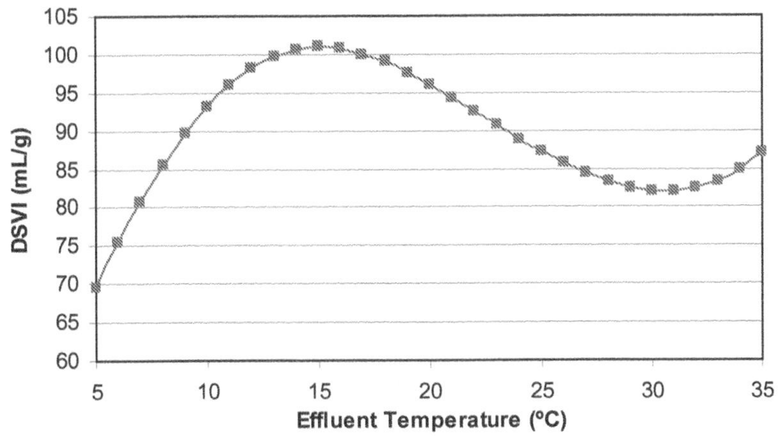

Figure 4-45b. Effluent Temperature vs. DSVI – Cubic Polynomial Regression Curve – Long-MCRT Plants with Selectors.

Summary

The long-MCRT WWTP regression analyses results show that DSVI is lowest when the selector and ICZ HRTs and the selector volume to total basin volume ratio → 0, suggesting that systems not equipped with selectors yield the lowest DSVIs. The 7-d average selector MCRT does not significantly influence DSVI ($R^2 = 0.1\%$), nor does the total selector F/M ($R^2 = 0.2\%$) or activated sludge influent BOD_5/TSS ratio (raCOD is not a significant factor in long-MCRT bulking). This is in contrast to the regression analysis results for the short-MCRT plants with anoxic or anaerobic selectors group. These results suggest that an unaerated selector does not significantly control long-MCRT filamentous organisms and bulking, which is supported by the literature (Wanner, 1993; Jenkins et al., 2004; Martins et al., 2004b). Similar to the regression analysis results for the short-MCRT unaerated selector group, however, the long-MCRT plant regression analysis showed that the ICZ F/M did not significantly influence the DSVI ($R^2 = 0.1\%$). This suggests that kinetics are not important to control either short- or long-MCRT filamentous organisms in unaerated selectors.

Table 4-17 summarizes the best ranges found in this study (as shown by the regression analysis) and the literature for the most important design and operating parameters for controlling filamentous bulking in long-MCRT activated sludge systems.

Table 4-17. Recommended Parameter Ranges for Long-MCRT Plants with Selectors.

Parameter	Recommendations from this Study[1]	Recommendations from Literature	Literature References
Average MLSS (mg/L)	2,500-4,500+		
Selector HRT (with recycle) (h)	→ 0	0.75-2.0	Jenkins et al., 2004
ICZ HRT (with recycle) (min)	→ 0		
Selector Vol/Total Basin Vol.Ratio	→ 0	25	Wanner, 1994
Number of Aeration Basin Stages	more is better, up to 8	many	Jenkins et al., 2004
Selector HRT (without recycle) (h)	→ 0		
Number of Selector Stages	0	3	Jenkins et al., 2004; Wanner, 1994
Effluent pH	6.4-6.7 best (7.7+ worst)		
Effluent Temperature (°C)	27-32 best (13-17 worst)		
% RAS Flow (%)	not significant	≤100	Wanner, 1994
Activated Sludge Influent BOD (mg/L)	not significant		
Aeration Basin DO (mg/L)	not significant	>1-2	Jenkins et al., 2004; Wanner, 1994
ICZ HRT (without recycle) (h)	not significant		
Reactor MCRT (d)	not significant		
Selector F/M [kg BOD_5/(kg MLSS·d)]	not significant	≤1.0	Jenkins et al., 2004
BOD/TSS Ratio	not significant		
Selector MCRT (d)	not significant	1-2	Jenkins et al., 2004
ICZ F/M [kg BOD_5/(kg MLSS·d)]	not significant	6	Wanner, 1994

Note:

[1] "→ 0" denotes "approaching zero." Although the regression line or curve shows the parameter is zero when the DSVI is lowest, the data in this study does not have parameter values equal to zero.

4.4.13 Percentile Distribution Analysis

An analysis of percentile distributions for selected parameters in each of the three datasets (short-MCRT plants with aerobic selectors, short-MCRT plants with anoxic or anaerobic selectors, long-MCRT plants with selectors) was conducted (see Appendix E) to further evaluate data variation within each plant category. The analysis yielded the following information:

- DSVIs for the long-MCRT plants were significantly lower than the short-MCRT plants; SVI values did not vary as significantly between datasets. Wanner (1994) suggested that this is because filamentous bacteria that grow at long MCRTs produce lower SVIs.
- ICZ F/M values were significantly higher and varied over a broader range in the short-MCRT plants with aerobic selectors, and long-MCRT plants had the lowest ICZ F/Ms.
- The activated sludge influent BOD_5 concentration was significantly higher in the long-MCRT plants and varied over a broader range. Activated sludge influent BOD_5/TSS ratios were higher and more variable in the long-MCRT plants and short-MCRT plants with anoxic and anaerobic selectors than in the short-MCRT plants with aerobic selectors.
- Selector F/Ms were lowest in the long-MCRT plants with significant variation, while selector HRTs were highest and highly variable within this plant group.
- Aeration basin DO concentration was higher in the long-MCRT plants than the short-MCRT plants with anoxic or anaerobic selectors.

4.4.14 Computerized Selector Diagnostic Tool

A computerized selector diagnostic tool was prepared as part of this project so that the regression analysis results from this study could be easily used to assist those troubleshooting or designing a selector installation. The computerized selector diagnostic tool is provided on a CD-ROM located on the inside back cover of this report. Documentation for this software is provided in Appendix F.

4.5 Conclusions

The analysis of average selector operating data versus 90th percentile SVI and DSVI yielded the following main conclusions:

- Anoxic selector installations appear to provide superior settleability control compared to anaerobic selectors. Approximately 85% of anoxic selector plants (23 of 27) had 90th percentile DSVIs $\leq$150 mL/g, while only 14% of anaerobic plants (two of 14) achieved this result. Most anoxic selector installations, however, were in long-MCRT plants, while all anaerobic selectors were at short-MCRT plants. The lower DSVI in plants with anoxic selectors may be due to the types of filamentous bacteria that grow at long MCRTs (Wanner, 1994) versus short MCRTs rather than the selector type.
- Selector staging did not have a significant impact on settleability in anoxic selector systems. In fact, all eight single-stage anoxic selectors yielded 90th percentile DSVIs $\leq$150 mL/g, while four of 18 multi-stage anoxic selectors exceeded this value.
- Selector staging did not have a significant impact on settleability in anaerobic selector systems, since six of seven systems yielded 90th percentile DSVIs of >150 mL/g in both the single- and multi-stage categories.

- Based on plots of average values vs. 90th percentile DSVIs, no significant relationships could be found between settleability control and selector ICZ F/M, selector F/M, selector MCRT, system MCRT (excluding clarifier solids), contact loading, or selector HRT.

Comparing average parameter and 90th percentile SVI/DSVI values for the plants included in the detailed plant investigation is somewhat limited since each facility is represented by only a single data point and does not reflect variation in each parameter. A single-variable regression analysis, incorporating daily operating data for each facility, was conducted to better evaluate the influence of parameter variation on SVI and DSVI values.

The regression analysis for short-MCRT plants with anoxic or anaerobic selectors, short-MCRT plants with aerobic selectors, and long-MCRT plants with selectors yielded the following main conclusions:

- For short-MCRT plants, anoxic or anaerobic selectors should be sized large enough to remove all or most of the raCOD and should be staged to prevent short-circuiting and raCOD breakthrough to the main aeration basin rather than to provide a kinetic advantage.
- For short-MCRT plants with aerobic selectors, a substrate concentration gradient should be provided to give a kinetic advantage to floc-formers over filamentous organisms; however, at higher influent BOD_5 concentrations, sufficient raCOD may leak through to the main aeration zone to cause bulking problems.
- Selectors do not significantly control filamentous organisms and bulking in long-MCRT plants, as also indicated in the literature (Wanner, 1993; Jenkins et al., 2004; Martins et al., 2004b).

CHAPTER 5.0

FULL-SCALE DEMONSTRATION PROJECTS

5.1 Introduction

This study included full-scale anaerobic selector demonstration studies at two wastewater treatment facilities—the East Bay Municipal Utility District (EBMUD) Main Wastewater Treatment Plant (MWWTP) in Oakland, Calif., and the Orange County Sanitation District (OCSD) Plant No. 1 in Fountain Valley, Calif. The goal of this work was to provide municipalities with key information necessary for successful selector implementation at their facilities by highlighting process considerations and issues. More importantly, however, this full-scale study demonstrated how the recommended design/operating ranges for significant parameters presented in Chapter 4.0 can be used to explain selector performance. This section presents the main findings and conclusions of the two full-scale anaerobic selector demonstration projects conducted by EBMUD and OCSD.

5.2 East Bay Municipal Utility District Main Wastewater Treatment Plant

5.2.1 Background

The EBMUD MWWTP is a high-purity oxygen activated sludge (HPOAS) plant with an annual average daily flow of 80 MGD. Figure 5-1 is a percentile distribution of SVI measurements from January 1999 to December 2002. The plot indicates that a typical SVI control limit of 150 mL/g was achieved only about 25% of the time. Although plant operations personnel are required to implement RAS chlorination when the SVI reaches 200 mL/g, SVI levels exceed 300 mL/g nearly 10% of the time.

Figure 5-1 also indicates that the dominant filaments responsible for causing sludge bulking at the MWWTP are Type 1701, Type 021N, and *S. natans*. The data is based on microscopic analyses of mixed liquor samples conducted by EBMUD laboratory staff from January 1999 to December 2002, using a qualitative scale range from 0 (none) to 6 (excessive) to rank individual filament types. A filament was classified as "dominant" if a 5 (abundant) or 6 (excessive) abundance level was identified. The average dominant result is also plotted for each filament type, indicating that Type 021N tended to have the highest frequency of excessive results when dominant. Although not a bulking filament, nocardioforms were reported as a dominant filament in nearly 50% of all samples, which explains the significant foaming problems commonly experienced at the MWWTP. In an effort to improve secondary process control and reliability, EBMUD conducted a bench- and full-scale evaluation of the anaerobic selector process from April to September 2001 and from June to October 2003, respectively.

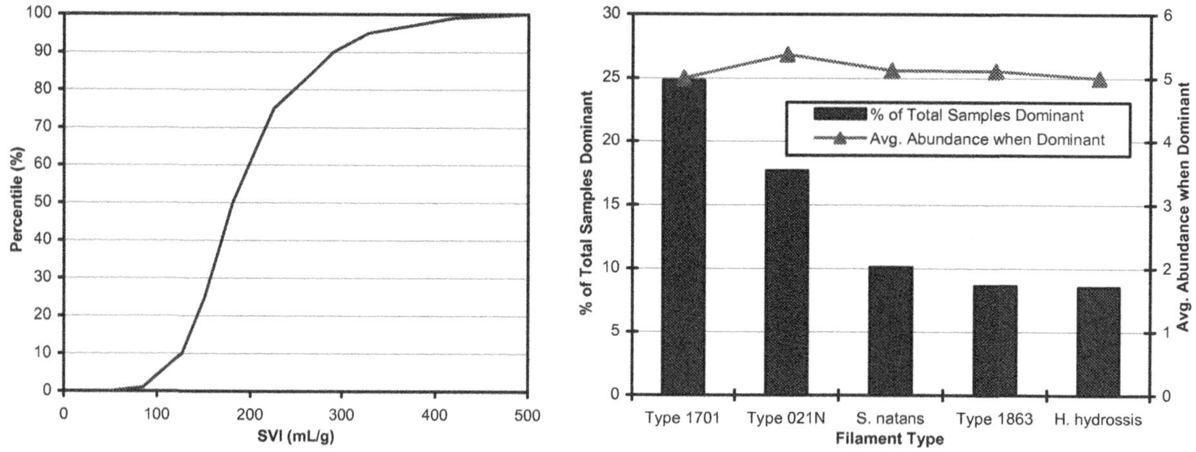

Figure 5-1. EBMUD SVI Percentile Distribution and Dominant Filament Results (1999-2002).

5.2.2 System Description

The HPOAS secondary treatment process at the MWWTP originally consisted of eight aeration basins (1.6 MG each), each divided into four equal volume stages (0.4 MG each, 46 ft x 46 ft x 25 ft deep), and 12 secondary clarifiers (surface area = 15,400 ft^2, 140 ft dia., 14 ft deep). Aeration high purity oxygen was provided by 100-hp surface aerators in Stages 1 and 2, and 50-hp surface aerators in Stages 3 and 4. The system was designed to provide step-feed with the secondary influent flow split evenly between Stages 1 and 2. The MWWTP may be operated in split-plant mode with two different aeration basin and secondary clarifier configurations. Initial concepts on implementing an anaerobic selector at the MWWTP focused on converting the first stage of each aeration basin to an anaerobic zone (25% of the total reactor volume). As an initial step, EBMUD decided to conduct a bench-scale evaluation to determine whether this configuration would provide effective bulking control. Important considerations included whether an active PAO population could be established at MCRTs typical of MWWTP operation (1–2 d) and whether the first stage was adequately sized for effective anaerobic selector operation.

5.2.3 Bench-Scale Anaerobic Selector Evaluation

EBMUD conducted a bench-scale anaerobic selector evaluation from April to September 2001. Two 8-L CMAS reactors were operated in parallel with one equipped with a selector and the other serving as the control. The selector system included a 2-L selector and a 6-L aeration zone to match the full-scale MWWTP configuration. The systems were fed primary effluent collected daily from the MWWTP.

Since Type 1701 and Type 021N were problematic at full-scale, initial objectives included promoting the growth of these filaments in both reactor systems; however, these efforts, which included adding an additional source of raCOD[1] to the influent feed (acetate), were largely unsuccessful, as *S. natans* was the dominant filament type. *S. natans* persists in bench-scale units due to the inability to control seeding from biological growth on tubing and reactor walls (Gabb et al., 1989). Low abundance levels for both Type 1701 and Type 021N were present throughout the evaluation. In addition, since physical foam trapping issues common in most HPOAS systems were not present, nocardioform filament levels were also low.

[1] Refer to raCOD discussion on Page 1-3 in Chapter 1.0.

Efforts to develop an active PAO population at bench-scale at total reactor MCRTs of 1.0, 2.0, and 3.0 d were unsuccessful. No significant orthophosphate release occurred in the anaerobic selector zone. This condition required seeding both reactor systems with an active PAO population from a nearby full-scale treatment facility equipped with an anaerobic selector (Central Contra Costa Sanitary District, Martinez, Calif.). Following seeding, the selector system maintained an active PAO population for the duration of the experiment (7 MCRTs), with an average orthophosphate concentration of 22.0 mg P/L in the selector zone compared to 5.5 mg P/L in the control system. The presence of PAOs and a "selector effect" yielded an average DSVI of 210 mL/g in the selector system, while the average DSVI in the control was 710 mL/g due primarily to *S. natans*. A summary of the control and selector CMAS reactor operating data following seeding is presented in Table 5-1.

Table 5-1. EBMUD Bench-Scale Anaerobic Selector Evaluation Results (MCRT = 3.0 d).

Parameter	Primary Effluent	Selector Mixed Liquor		Control Mixed Liquor	Secondary Effluent	
		Anaerobic	Aerobic		Selector	Control
COD (mg/L)	640	--	--	--	165	145
fCOD (mg/L)	350	--	--	--	105	110
TSS (mg/L)	170	--	2,600	1,800	20	50
Ortho-P (mg P/L)	4.5	22.0	6.4	5.5	2.4	3.3
Diluted SVI (mL/g)	--	--	210	710	--	--
S. natans Abundance	--	--	3.6	6.0	--	--

Figure 5-2 is a plot of DSVI levels in both the selector and control CMAS systems following seeding at a total reactor MCRT of 3.0 d.

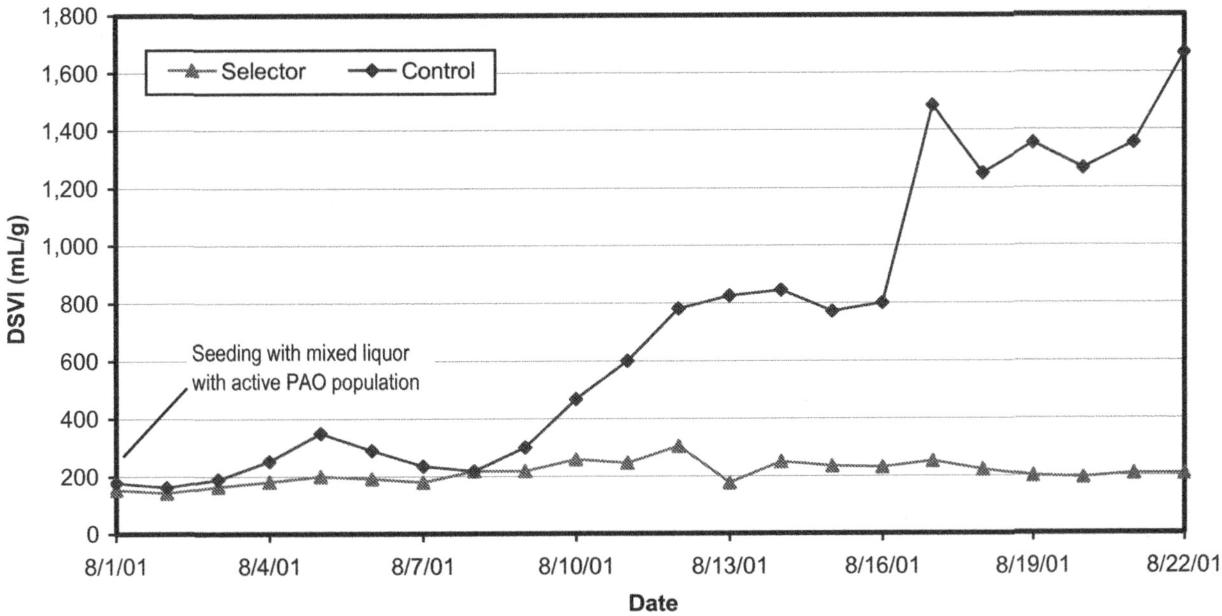

Figure 5-2. EBMUD Bench-Scale Selector and Control DSVI (following seeding) at MCRT = 3.0 d.

Although difficulties were encountered in demonstrating the control of specific filaments (Type 1701 and Type 021N) and developing an active PAO population without seeding, the bench-scale evaluation provided the following conclusions:

- An anaerobic selector compartment sized at 25% of the total reactor volume can provide effective SVI control;
- An anaerobic selector is able to control the growth of *S. natans* relative to a control.
- Once developed, an active PAO population may be maintained at a total reactor MCRT of 3.0 d.

Based on the results of the bench-scale study and the need to improve secondary process control and reliability, EBMUD decided to implement a full-scale pilot evaluation of the anaerobic selector process.

5.2.4 Full-Scale Selector Process Modifications

In August 2002, EBMUD completed the construction modifications required for a partial anaerobic selector conversion. The goal was to convert the first stage of two aeration basins to anaerobic selectors and to compare process performance to the remaining basins under split-plant mode operating conditions. The required process modifications included replacing the 100-hp Stage 1 surface aerators on Aeration Basins 1 and 2 with 25-hp submerged mixers, and converting secondary influent feed from step-feed to plug-flow mode. The selector conversion did not include relocation of the pure oxygen feed lines to Stage 1 and sealing off the anaerobic zone from the flow of pure oxygen between interstitial openings in the walls separating each stage. Although pure oxygen was present in the headspace above the Stage 1 selectors, oxidation reduction potential (-300 to -250 mV) and dissolved oxygen (DO <0.5 mg/L) measurements indicated the presence of anaerobic conditions in these zones.

5.2.5 Selector Design Criteria

The initial selector design and operating criteria is presented in Table 5-2. The anaerobic selector was sized at a nominal HRT of 35 min, which is low relative to the recommended range of 45–120 min (Jenkins, 2004).

Table 5-2. Summary of Initial EBMUD MWWTP Anaerobic Selector Design and Operating Criteria.

Parameter	Value
Number of Selector Stages	1
Selector HRT, nominal (min)	35
Selector F/M [kg $cBOD_5$/(kg VSS·d)]	3.6
Total System F/M [kg $cBOD_5$/(kg VSS·d)]	0.9
Selector MCRT (d)	0.7
Aerated MCRT (d)	2.0
Reactor MCRT (d)[1]	2.7
MLSS (mg/L)	2,000
Selector DO (mg/L)	< 0.2
Selector Oxidation-Reduction Potential (mV)	-300 to -100

Notes: [1] Excludes secondary clarifier solids.

5.2.6 Results and Discussion

The MWWTP was operated in split-plant mode from June to October 2003. The selector and control performance and operating data is presented in Table 5-3. Average selector and control SVIs during the evaluation were 120 and 270 mL/g, respectively. Filament abundance data indicated that Type 021N was present and frequently dominant (5 or 6 on abundance scale) in both systems, while the selector appeared to provide some control of *S. natans*. The selector and control plants were operated at an average aerated MCRT (excluding clarifier solids) of 1.0

and 0.6 d, respectively. Selector and control SVIs, aerated MCRTs, and MLSS concentrations are plotted in Figure 5-3.

Table 5-3. EBMUD MWWTP Anaerobic Selector Performance and Operating Data (June 12 – October 31, 2003).

Parameter	Selector	Control
Flow (MGD)	34.3	32.2
Selector F/M [kg BOD_5/(kg MLSS·d)][1]	5.1	N/A
Total System F/M [kg BOD_5/(kg MLSS·d)][1]	1.3	2.5
Selector MCRT (d)	0.3	N/A
Aerated MCRT (d)	1.0	0.6
Reactor MCRT (d)[2]	1.3	1.0
Avg. Selector HRT (w/RAS/w/o RAS) (min)	26/34	N/A
Activ. Sludge Influent BOD_5/TSS Ratio[1]	2.4	2.4
MLSS (mg/L)	2,040	940
SVI (mL/g)		
Average	120	270
90th Percentile	166	471
Effluent TSS (mg/L)	12	18
Effluent $cBOD_5$ (mg/L)	9	15
Orthophosphate (mg-P/L)		
Secondary Influent	4.1	4.1
Stage 1	12.0	3.3
Stage 4	1.3	2.9
Filament Abundance[3]		
Type 021N	4.1 (33)	3.8 (33)
Type 1701	2.7 (0)	2.9 (0)
S. natans	2.5 (0)	3.7 (16)
Nocardioform	4.0 (23)	2.5 (11)
Nocardioform Count (10^6 filament intersections/g VSS)	0.9	0.4

Notes:

[1] Value reported on a $cBOD_5$ basis and converted to BOD_5 using $BOD_5 = 1.45 \times cBOD_5$.

[2] Excludes secondary clarifier solids.

[3] Values shown in parentheses represent total number of dominant samples (5 or 6 on abundance scale) for a given filament type during the study period, per Jenkins *et. al.*, 2004.

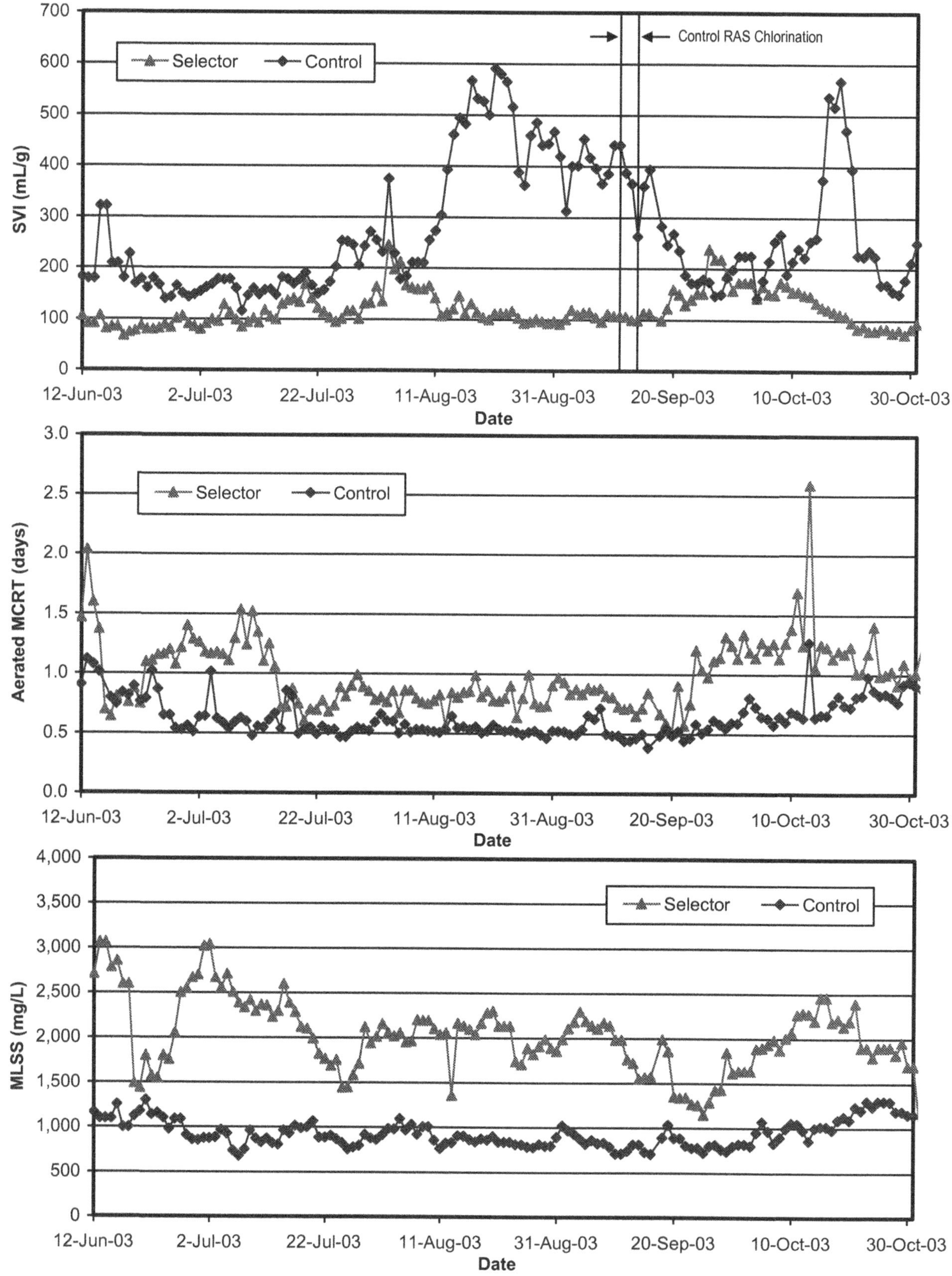

Figure 5-3. EBMUD MWWTP Full-Scale Selector and Control SVI, Aerated MCRT, MLSS (June 12 – October 31, 2003).

Although the control plant was intended to provide a basis for comparing the results achieved in the selector plant, the two plants were not operated under the same test conditions. The selector plant was operated at a higher target MCRT to promote the growth and proliferation of PAOs, while the control plant was operated under normal conditions (aerated MCRT in the range of 0.5–1.0 d) primarily to avoid problems with excessive nocardioform foaming. Attempting to operate the selector plant at a higher MCRT caused excessive foaming problems primarily on the mixed liquor channels. Increased foaming in the selector plant is supported by the nocardioform count and abundance summary data provided in Table 5-3 and the plot of nocardioform count presented in Figure 5-4 below. Historically, the MWWTP has had significant issues with foaming control, and the higher MCRT required for the anaerobic selector process exacerbated these foaming problems. Operations personnel attempted to operate the plant at MCRTs that satisfied both the higher MCRT requirement for selector operation and the lower MCRT requirement for nocardioform foaming control but had limited success. Attempts to increase the aerated MCRT in mid-June and mid-July 2003 (refer to Figure 5-3) resulted in foaming episodes, which required an increase in wasting rate and a corresponding drop in MLSS.

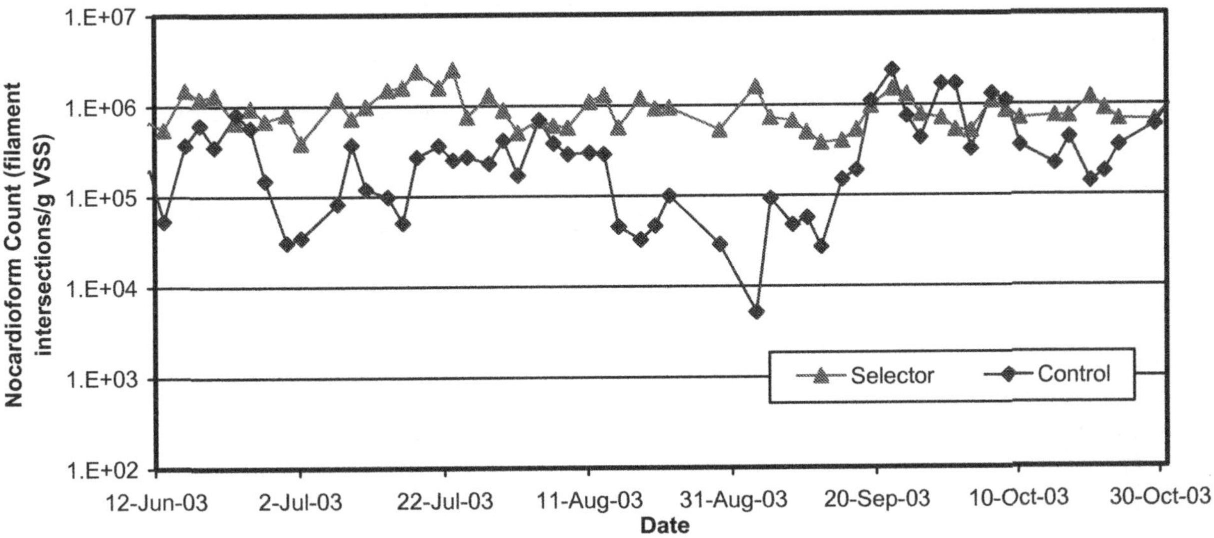

Figure 5-4. EBMUD MWWTP Full-Scale Selector and Control *Nocardia* Counts (June 12 – October 31, 2005).

Table 5-3 indicates that there was a moderate release and uptake of orthophosphate levels in the selector plant relative to the control. Average orthophosphate levels for the selector plant in Stages 1 and 4 were 12.0 and 1.3 mg-P/L, respectively. Based on an average secondary influent orthophosphate level of 4.1 mg-P/L, approximately 7.9 mg-P/L of release occurred with an average uptake of 2.8 mg-P/L. In contrast, average orthophosphate levels for the control plant in Stages 1 and 4 were 3.3 and 2.9 mg-P/L, respectively. Secondary influent VFA levels were also measured during the study as an indication of raCOD loading to the selector and control plants. Orthophosphate profiles and secondary influent VFAs for the selector and control plants are presented in Figures 5-5 and 5-6, respectively.

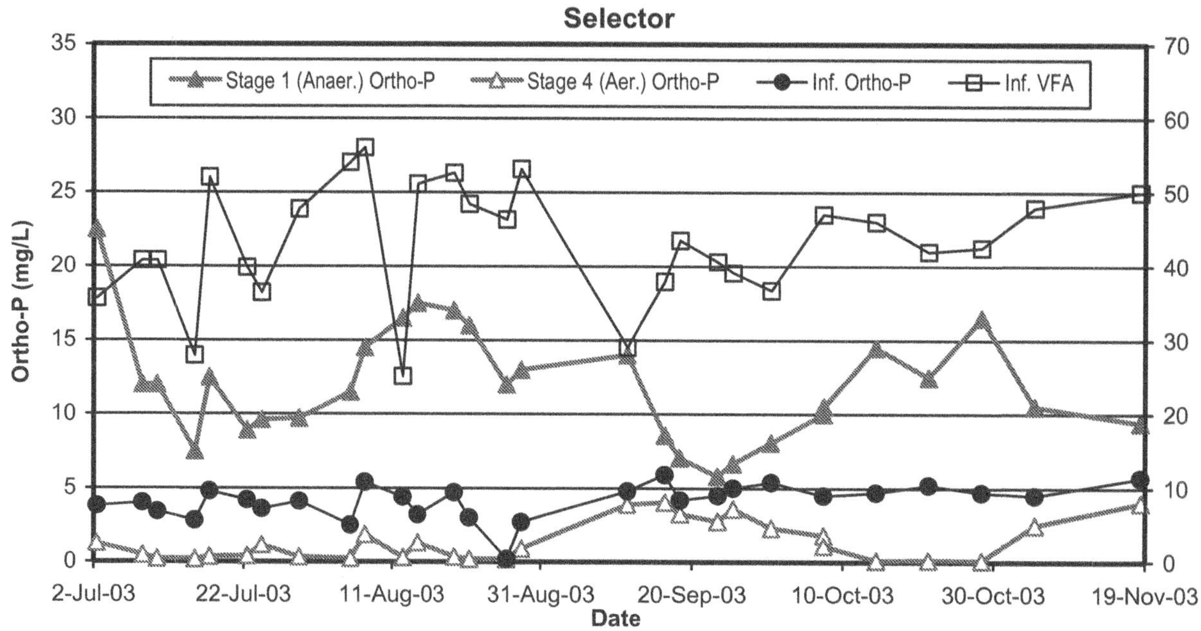

Figure 5-5. EBMUD MWWTP Full-Scale Selector Ortho-P Release and Uptake, Influent Volatile Fatty Acids (VFAs).

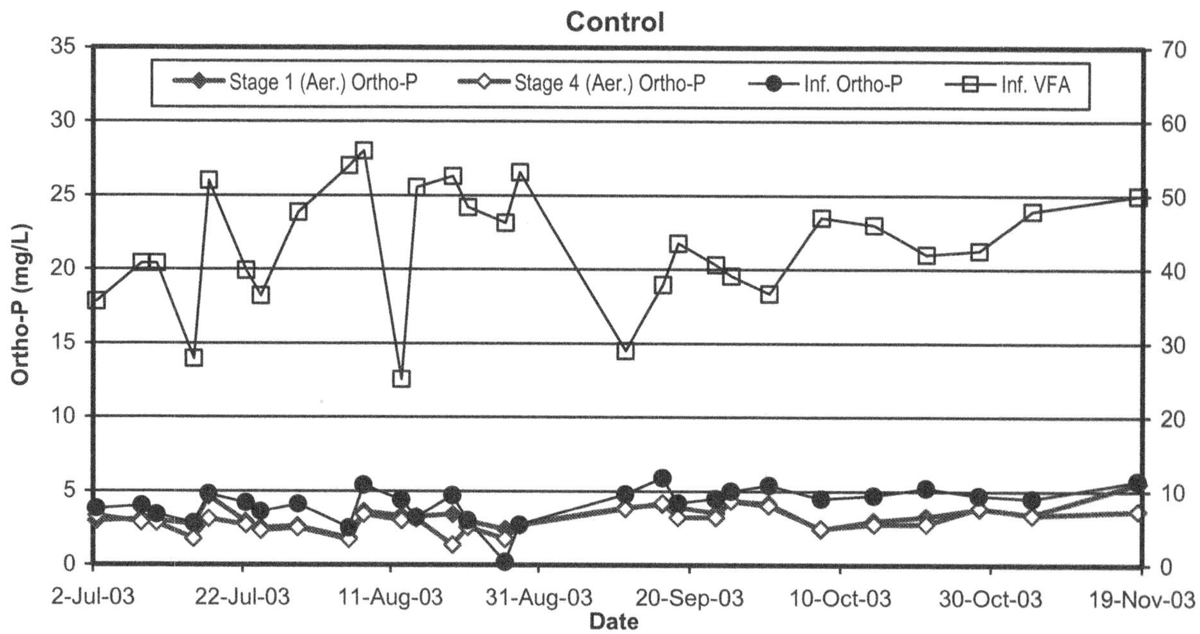

Figure 5-6. EBMUD MWWTP Full-Scale Control Ortho-P Levels, Influent VFAs.

5.2.7 Conclusions

The full-scale anaerobic selector evaluation conducted at EBMUD's MWWTP provided the following main conclusions:

- Installation of a single-stage anaerobic selector (nominal HRT = 34 min) provided significantly improved SVI control (average = 120 mL/g) relative to the control plant (average = 270 mL/g).

- The anaerobic selector demonstrated some control of Type 1863 and *S. natans*; however, Type 021N persisted in the selector plant at nearly the same abundance as the control.
- Operation at an aeration MCRT of 1.0 d was able to maintain an active PAO population in the selector plant with an anaerobic zone P release of approximately 8 mg-P/L.
- Significant nocardioform foaming problems were associated with the increased MCRT required for the anaerobic selector process. The anaerobic selector did not provide any nocardioform foaming control benefits.

Based on the results of the full-scale pilot evaluation, EBMUD decided to move forward with conversion of the six remaining aeration basins to anaerobic selector systems. The construction modifications required for this work were completed in July 2005.

5.3 Orange County Sanitation District Plant No. 1

5.3.1 Background

Historically, OCSD Plant No. 1 has been operated at a low MCRT with a high SVI to produce a high quality, low turbidity plant effluent suitable for use at the OCSD groundwater reclamation facility. Plant staff has identified the presence of both low DO filamentous organisms (*S. natans*, and Type 1701) and sulfide oxidizing filaments (*Thiothrix*, Type 021N), which typically produced SVIs ranging from 300–600 mL/g at Plant No. 1. In 2004, OCSD began evaluating options to allow expansion of the secondary treatment capacity at Plant No. 1, including installation of an anaerobic selector to reduce SVIs and associated secondary clarifier capacity requirements. In July 2004, OCSD completed construction modifications to equip half of the aeration basins at Plant No. 1 with anaerobic selectors and began a full-scale pilot evaluation.

5.3.2 System Description

Plant No. 1 is an air activated sludge plant with 10 aeration basins (1.4 MG each, 275 ft long x 45 ft wide x 15 ft deep) and 24 secondary clarifiers (surface area = 21,000 ft^2, 150 ft long x 40 ft wide x 10 ft deep). Each aeration basin is divided into six equal-volume stages with a higher air diffuser density in the first four stages. Similarly to EBMUD's MWWTP, Plant No. 1 may be operated in a split-plant mode, which divides the facility into two plants with separate sludge recirculation and wasting control—one plant consists of Aeration Basins 1–5 and the odd-numbered clarifiers, while the second plant consists of Aeration Basins 6–10 and the even-numbered clarifiers.

5.3.3 Selector Process Modifications

Stage 1 in each of Aeration Basins 1–5 was converted into anaerobic selectors by installing subsurface mixers, shutting off the air diffusers, and improving the existing baffling between Stages 1 and 2. No modifications were made to Aeration Basins 6–10. This process configuration allowed Plant No. 1 to be operated in split-plant mode with Aeration Basins 1–5 operating as the selector and 6–10 serving as the control.

5.3.4 Selector Design Criteria

The initial anaerobic selector design and operating criteria are presented in Table 5-4. The selector was sized for a nominal HRT of 45 min, which is at the lower end of the recommended range of 45–120 min (Jenkins, 2004). Although anaerobic selectors are typically operated at reactor MCRTs (excluding clarifier solids) greater than 1.8 d to promote the growth of PAOs, a lower initial MCRT was selected as the starting point to determine the relationship between MCRT and PAO activity.

Table 5-4. Summary of Initial OCSD Plant No. 1 Anaerobic Selector Design and Operating Criteria.

Parameter	Value
Number of Selector Stages	1
Selector HRT, nominal (min)	45
Selector F/M [kg BOD_5/(kg VSS·d)]	6.0–8.4
Total System F/M [kg BOD_5/(kg VSS·d)]	1.0–1.4
Selector MCRT (d)	0.17–0.21
Aerated MCRT (d)	0.85–1.05
Reactor MCRT (d)[1]	1.0–1.2
MLSS (mg/L)	600–700
Selector DO (mg/L)	<0.2
Selector Oxidation-Reduction Potential (mV)	-300 to -100

Notes: [1] Excludes secondary clarifier solids.

5.3.5 Results and Discussion

Plant No. 1 was operated in split-plant mode from July to November 2004 to allow a full-scale evaluation of the anaerobic selector process. The evaluation was divided into four phases based primarily on the target MCRT. The selector performance and operating data for all four phases is provided in Table 5-5. Selector and control SVI, MCRT, and Stage 1 orthophosphate levels are presented in Figures 5-7, 5-8, and 5-9, respectively.

Table 5-5. OCSD Plant No. 1 Anaerobic Selector Performance and Operating Data.

Parameter	Phase 1 7/20–8/16/04		Phase 2 8/17–9/11/04		Phase 3 9/12–10/21/04		Phase 4 10/22–11/20/04	
	Selector	Control	Selector	Control	Selector	Control	Selector	Control
Flow (mgd)	30.5	34.3	31.3	36.3	29.1	30.8	24.5	31.1
Selector F/M [kg BOD_5/(kg MLSS·d)][1]	8.1	N/A	6.3	N/A	4.6	N/A	4.3	N/A
Total System F/M [kg BOD_5/(kg MLSS·d)][1]	1.4	1.4	1.0	1.4	0.8	1.0	0.7	0.9
Selector MCRT (d)	0.21	N/A	0.26	N/A	0.33	N/A	0.33	N/A
Aerated MCRT (d)	1.0	1.1	1.3	1.1	1.6	1.1	1.6	1.2
Reactor MCRT (d)[2]	1.2	1.1	1.6	1.1	1.9	1.1	1.9	1.2
Aeration Basin DO (mg/L)	2.0	2.0	1.9	1.8	1.5	1.3	1.4	1.1
Avg. Selector HRT (w/RAS/w/o RAS) (min)	31/54	N/A	31/52	N/A	32/57	N/A	38/67	N/A
Activ. Sludge Influent BOD_5/TSS Ratio[1]	1.8	1.7	1.7	1.6	2.1	2.0	2.4	2.3
MLSS (mg/L)	592	625	772	654	957	760	831	777
Selector DO (mg/L)	<0.2	-	<0.2	-	<0.1	-	<0.1	-
SVI (mL/g)	688	468	556	226	504	297	345	339
Turbidity (NTU)	5.0	3.9	4.2	3.5	4.3	4.2	4.3	4.0
Effluent TSS (mg/L)	9.0	8.2	8.0	6.2	8.7	8.1	9.1	7.9
1st Stage sCOD (mg/L)	70	60	77	72	75	63	69	68
1st Stage P (mg PO_4-P/L)	3.6	2.3	4.7	2.9	10.3	4.8	9.1	4.0

Notes: [1] Value reported on a COD basis and converted to BOD_5 using BOD_5 = 0.5 x COD.
[2] Excludes secondary clarifier solids.

During Phase 1 (July 20–August 16, 2004), both the selector and control plants were operated at normal process control target values (MCRT = 1.0–1.2 d) to determine whether an active PAO population could be established at a low MCRT. The selector plant activated sludge had higher SVIs than the control activated sludge, and there was no evidence of significant orthophosphate release in the anaerobic selector zone during Phase 1.

Phase 2 (August 17–September 11, 2004) began with an increase in the target MCRT from approximately 1.2 to 1.6 d to promote the growth of PAOs in the selector plant, while the control plant remained unchanged. An increased phosphorus release occurred in the selector and the SVI dropped, but the control plant SVI was still significantly lower (Table 5-5).

Although the target MCRT for the selector plant was raised further to 1.9 d during Phase 3 (September 12–October 21, 2004), only modest increases in orthophosphate release and reductions in SVI were observed (Table 5-5). The control plant continued to yield lower SVIs throughout Phases 1–3.

OCSD staff found that the selector plant was receiving approximately 30% less air flow because of the increased head loss resulting from shutting off the air diffusers to the first stage of each selector-equipped aeration basin. As a result, the DO levels in the first aerobic zone following the selector ranged from 0.1 to 0.7 mg/L. Two alternatives were proposed to prevent these conditions from promoting the growth of low DO filaments in the main aeration zone: 1) reduce the secondary influent flow rate to the selector plant, or 2) increase the air flow rate to the selector. Because it was extremely difficult to balance the air flows, wastewater flow to the selector plant was reduced and the flow to the control plant was increased during Phase 4 (October 22–November 20, 2004).

The air flow to Plant No. 1 is controlled by DO setpoints designed to prevent nitrification, so that as the MCRT was progressively raised, plant staff concurrently reduced the air supply to prevent nitrification. Table 5.5 shows that the aeration basin DO dropped as the MCRT was increased from Phase 1 to Phase 4 of the study. Although selector performance improved, the SVI values were still higher than the typical process control limit of 150 mL/g.

The selector plant experienced severe filamentous bulking throughout all four test phases (Figure 5-7). The dominant filaments identified were Type 021N, *Thiothrix*, Type 1701, and *S. natans*, which can be controlled by a selector process. Intracellular sulfur granules were present in the Type 021N and *Thiothrix* filaments. Selector zone effluent soluble COD remained above the target value of 60 mg/L (Jenkins *et. al.*, 2004), and phophorus release in the selector was minimal throughout the evaluation (Table 5-5).

Figure 5-10 is a summary of the SVIs and anaerobic zone orthophosphate releases achieved as the MCRT was raised progressively from 1.2–2.0 d. The data suggests that an MCRT >2.0 d may be necessary to achieve a "selector effect," an active PAO population, and perhaps adequate SVI control. Since aeration basin DO dropped as the system MCRT was increased, this suggests that insufficient MCRT rather than low DO was responsible for poor selector performance.

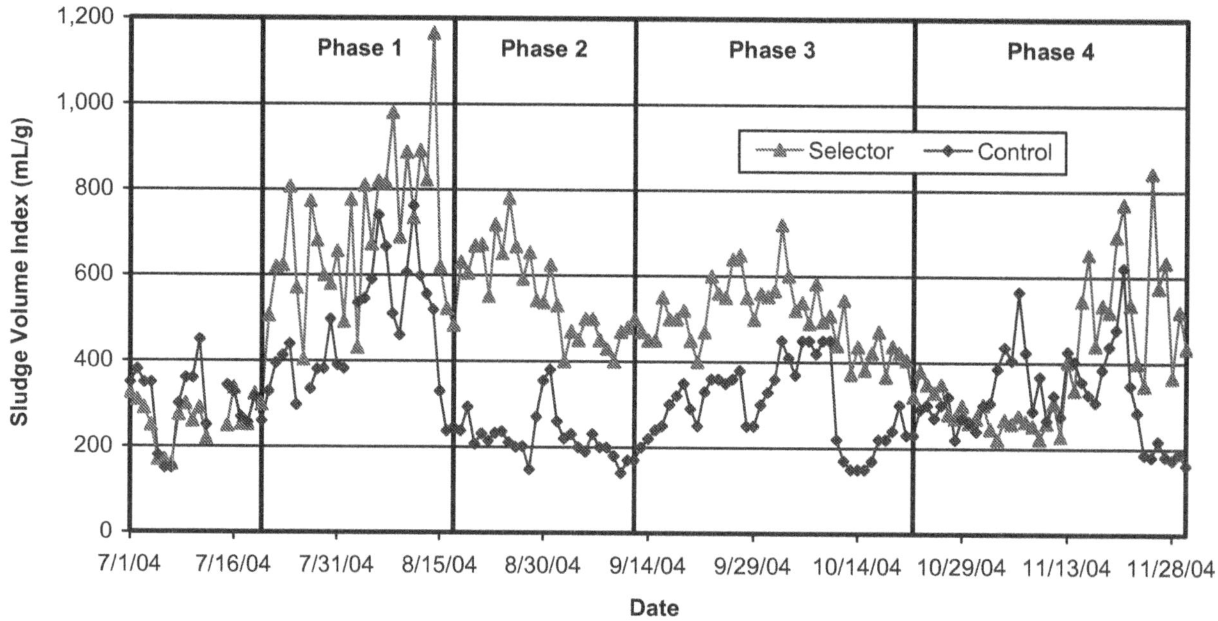

Figure 5-7. OCSD Selector and Control SVI.

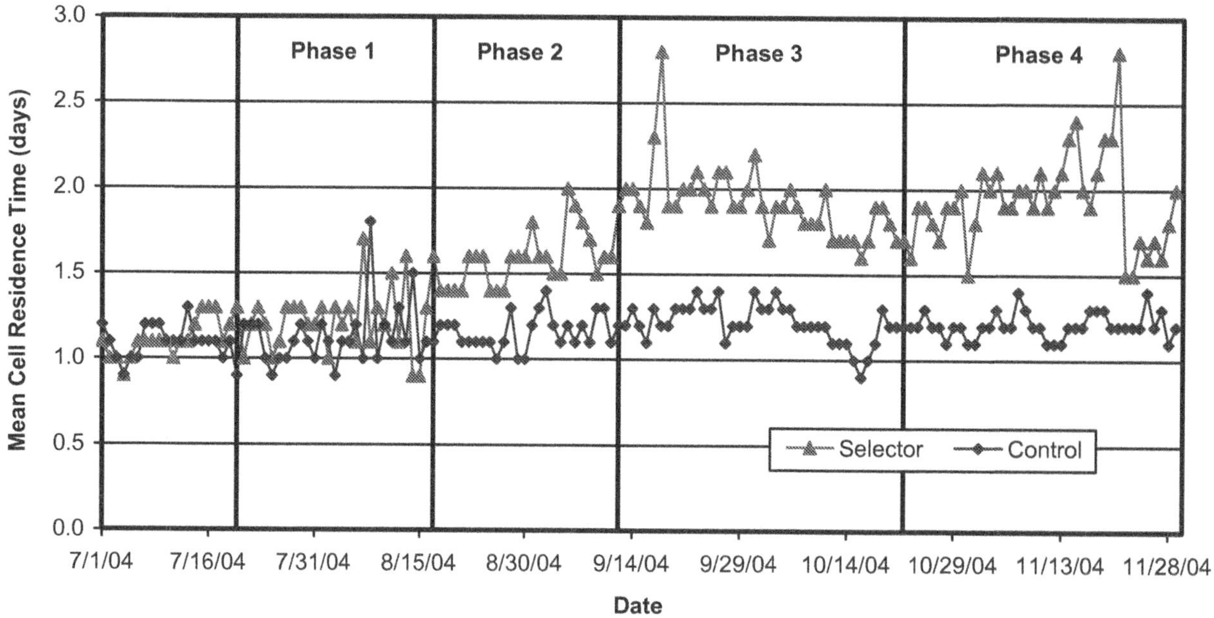

Figure 5-8. OCSD Selector and Control MCRT.

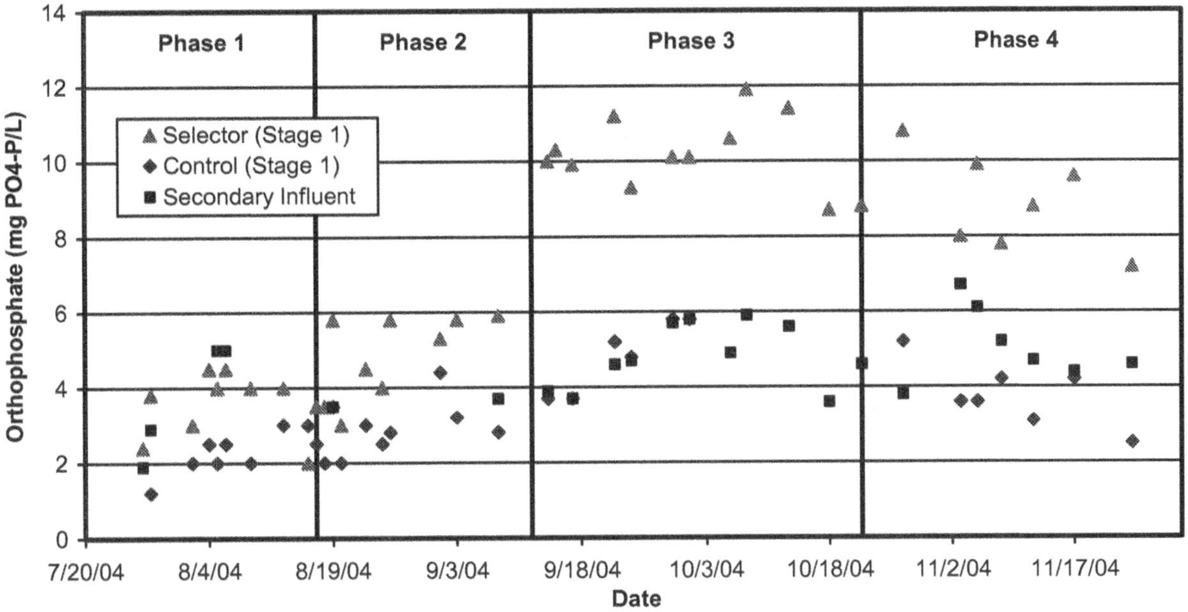

Figure 5-9. OCSD Selector and Control Stage 1 Orthophosphate Concentration.

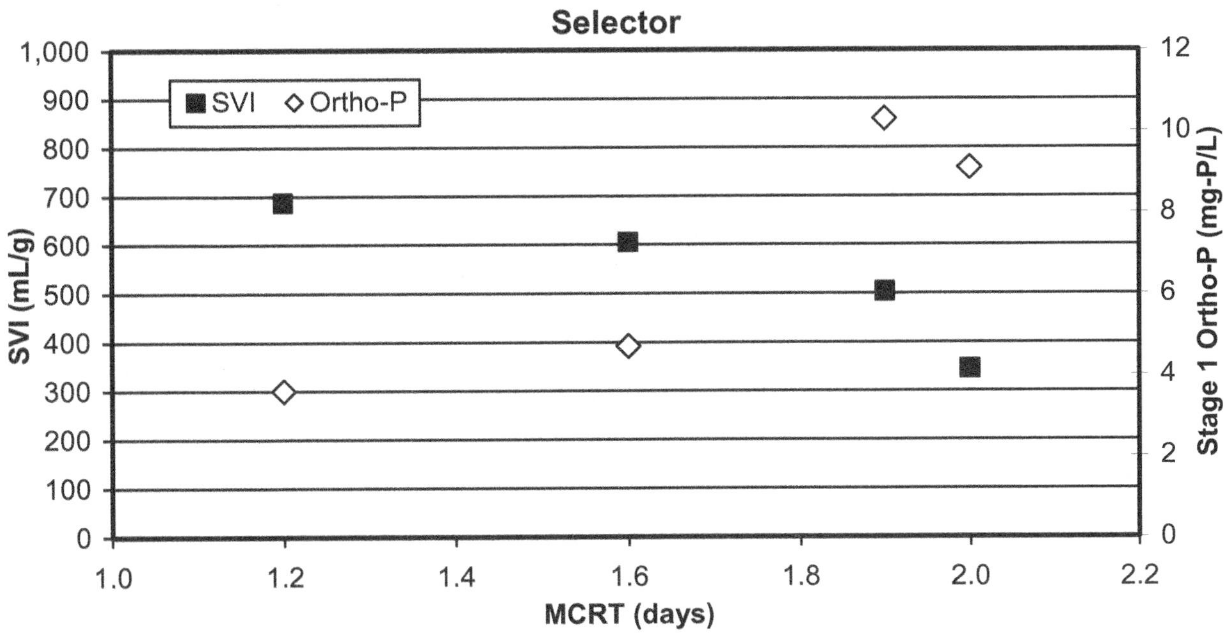

Figure 5-10. OCSD Reactor MCRT, SVI, and Stage 1 Orthophosphate Concentration.
Note: Graph is based on data provided by OCSD (Table 5-5).

5.3.6 Conclusions

The full-scale anaerobic selector evaluation conducted at OCSD Plant No. 1 yielded the following main conclusions:

- Installation of a single-stage anaerobic selector (nominal HRT = 45 min) resulted in significantly higher SVIs compared to the control over an MCRT range of 1.2–1.9 d. This may have been due to insufficient air flow and low DO conditions in the aeration basin and reduced aeration volume, since little or no phosphorus release (and therefore raCOD uptake) occurred in the selector.

- A significant PAO population could only be established at an MCRT >1.8 d.
- The presence of low DO and sulfide-oxidizing filaments in the selector system suggested that there was insufficient air feed, low DO conditions in the main aeration zone, and possible secondary influent septicity.

Based on the poor results of the full-scale evaluation, OCSD decided not to pursue installation of anaerobic selectors at Plant No. 1 as a method to increase secondary treatment capacity but may revisit selectors following future aeration system upgrades.

5.4 Recommendations for Conducting Selector Pilot Studies

Based on the lessons learned from the selector demonstrations conducted at EBMUD and OCSD, the following recommendations are made for conducting bench- and full-scale selector pilot studies:

Bench-Scale Studies

- Bench-scale selector studies should be conducted, whenever feasible, to verify selector sizing, loading, and configuration, demonstrate control of specific filaments of concern, and estimate required secondary operating parameters (MCRT, SVI, F/M, etc.).
- During bench-scale selector studies, special efforts should be made to control the growth of *S. natans* through daily cleaning and bleaching of system tubing and reactor walls.
- If initial attempts to develop an active PAO population in anaerobic selector experiments fail, it may be necessary to seed the reactors with mixed liquor from a full-scale plant equipped with an anaerobic selector. This does not necessarily mean that the full-scale plant will need to be seeded with PAOs.

Full-Scale Studies

- Whenever possible, the full-scale selector evaluation should be carried out in split-plant mode, similar to the EBMUD and OCSD evaluations, to allow comparison between the selector and a control plant.
- A specific recommendation regarding the target aerated MCRT necessary to support an active PAO is not supported by the data collected. It is likely that the required MCRT is site-specific due to variable orthophosphate release and uptake between facilities. Facilities may elect to begin the full-scale evaluation at a low MCRT and to progressively measure the impact on PAO activity and SVI. Orthophosphate levels in the secondary influent, anaerobic zone, and last oxic zone should be measured to assist in this evaluation.
- Mixed liquor samples should be analyzed for specific filament types and abundance to better understand the selector performance and limitations during the course of the study. This information is key to troubleshooting and making necessary process adjustments to the selector operating conditions.
- Special attention should be given to the aeration capacity in the initial oxic zone following the anaerobic zone to prevent low DO filamentous bulking episodes.

5.5 Comparison of EBMUD and OCSD Anaerobic Selector Performance

The full-scale pilot anaerobic selector evaluations conducted at EBMUD and OCSD yielded significant differences in terms of developing an active PAO population and demonstrating bulking control relative to a control plant. The EBMUD MWWTP selector plant average and 90th percentile SVIs were 120 and 166 mL/g, respectively, and were both lower than the control plant SVI values of 270 and 471 mL/g. Conversely, SVIs for the OCSD Plant No. 1 anaerobic selector system were higher than the control plant throughout most of the evaluation with no single reported SVI value <200 mL/g.

Although both the EBMUD and OCSD selectors were operated at about the same MCRT, significantly more orthophosphate release occurred in the EBMUD selector. Nonetheless, increasing OCSD's selector MCRT improved both orthophosphate release and SVI. Given the air supply problems encountered at OCSD, it is possible that low DO conditions in the initial oxic zone following the anaerobic selector aided in promoting the growth of filaments such as Type 1701. Since a sufficient PAO population was not developed to take up the available raCOD in the OCSD selector zone (per the low orthophosphate release), breakthrough to the main aeration zone may have allowed raCOD filaments, such as Type 021N and *S. natans* to predominate. OCSD also reported the presence of intracellular sulfur granules in both Type 021N and *Thiothrix* filaments, indicating the presence of secondary influent septicity. This condition may have been exacerbated when the anaerobic selector was installed.

Although the high SVI in the OCSD selector basins appeared to be caused by inadequate aeration, OCSD's SVI improved when the reactor MCRT was increased, even though the DO concomitantly dropped. Further, Table 5.5 shows that both aeration basin DO and SVI were lower in OCSD's control basins compared to their selector basins. A more thorough analysis is necessary to better understand why the OCSD selector failed while the EBMUD selector was successful. The differences between the EBMUD and OCSD selectors were further investigated by comparing the two selector systems' design/operating parameters to recommended ranges developed in this study and those ranges found in the literature. Table 5-6 presents this comparison.

Table 5-6 shows that the OCSD selector was operated outside of the range recommended by this study for all of the parameters considered. Similarly, the EBMUD selector was operated outside of the recommended range for all the parameters with the exception of MLSS and selector volume to total basin volume ratio, which were both optimum according to the recommended ranges. Although aeration basin DO concentrations were not measured at EBMUD during the evaluation, vent gas purities were between 70%–80^{+}% throughout the study period, which indicated that sufficient oxygen was fed to the aeration basins. Therefore, the EBMUD aeration basin DO was likely within the recommended range, and thus the EBMUD selector operated within this study's recommended ranges for the following parameters:

- Average MLSS
- Selector volume to total basin volume ratio
- Aeration basin DO concentration

Table 5-6. Comparison of EBMUD and OCSD Selector Operating and Performance Data.

Parameter	EBMUD	OCSD[1]	Recommendations from This Study	Literature Value	Literature Reference
Plant Type	high-purity O_2	air			
Selector Type	anaerobic	anaerobic			
Flow (MGD)	34.3	24.5			
SVI (mL/g)					
Average	120	345			
90th Percentile	166	536			
Orthophosphate (mg-P/L)					
Secondary Influent	4.1	5.1			
Stage 1 (anaerobic)	12.0	9.1			
Dominant Filament Types	Type 021N	Type 021N, *Thiothrix*, Type 1701, *S. natans*			
MLSS (mg/L)	2,040	831	1,500–2,000+		
Reactor MCRT (d)[2]	1.3	1.9	High as possible		
Selector F/M [BOD_5/(kg MLSS·d)]	5.1[3]	4.3[4]	<1.0 (lower is better)	≤1.0	Marten and Daigger (1997)
Selector MCRT (d)	0.32	0.32	2–3+	1–2	Marten and Daigger (1997)
Number of Selector Stages	1	1	2	3	Jenkins, 2004; Wanner, 1994
Aeration Basin DO (mg/L)	N/A	1.4	2.5–4.0 (air plants only)	>1-2	Jenkins et al. (2004)
Sec. Inf. BOD_5/TSS Ratio	2.4[3]	2.4[4]	<0.5		
Avg. Selector HRT (w/RAS/w/o RAS) (min)	26/34	38/67	>90/>150	45–120 (w/o recycle)	Jenkins et al. (2004)
Selector Volume to Total Basin Volume Ratio (%)	25	17	22.5-25.0	25	Wanner, 1994
Temperature (°C)	27	27	20-25 (27-30+ worst)		
Total System F/M [BOD_5/(kg MLSS·d)]	1.3[3]	0.7[4]	not significant		
Aerated MCRT (d)	1.0	1.6	not significant		

Notes: [1] Based on Phase 4 data.
[2] Excludes secondary clarifier solids.
[3] Value reported is on a $cBOD_5$ basis and converted to BOD_5 using $BOD_5 = 1.45 \times cBOD_5$.
[4] Value reported is on a COD basis and converted to BOD_5 using $BOD_5 = 0.5 \times COD$.

These parameters are discussed below relative to the EBMUD and OCSD selector studies.

Average MLSS

This study's regression analysis determined that average MLSS had the strongest influence on SVI, with an $R^2 = 22.4\%$ (see Table 4-12). The 7-day Reactor MCRT was next strongest with an $R^2 = 12.4\%$, or about half that of the average MLSS. The average MLSS in the EBMUD selector basins was 2,040 mg/L, or about 2.5 times as high as the average MLSS in the OCSD selector basins, which supports this study's regression results and provides a reason for the EBMUD selector's success and the OCSD selector's failure.

Selector Volume/Total Basin Volume Ratio

The EBMUD selector volume was 25% of the total basin volume, which is the optimum percentage according to this study's recommendations. In contrast, the OCSD selector was only 17% of the total basin volume and below the recommended range.

Aeration Basin DO

Based on high vent gas purities, the EBMUD aeration basins had sufficient oxygen during the selector study. The OCSD aeration basin DO, however, was only 1.4 mg/L in Phase 4 of the study but as high as 2.0 mg/L in Phase 1 of the study. These DO values are sufficient according to literature recommendations but are below the 2.5-4.0 mg/L recommended by this study. This agrees with the earlier assumption that OCSD's inadequate aeration contributed to its high SVIs during the selector study, but it also shows that low DO concentration was probably not the only cause for OCSD's high SVIs.

The EBMUD selector released about 8 mg/L phosphorus, compared to only 4 mg/L phosphorus for the OCSD selector, and achieved significantly lower SVIs. According to this study's regression analysis, the difference between EBMUD's and OCSD's selector performances was that EBMUD's selector system was operated within the recommended range for three key operating parameters, while the OCSD selector was not operated within any of the recommended ranges identified in this study.

The EBMUD selector achieved low SVIs, despite operating outside the recommended ranges from this study for the following significant parameters:

- Reactor MCRT
- Selector F/M
- Activated Sludge Influent BOD_5/TSS ratio
- Reactor F/M
- Selector MCRT
- Average Selector HRT (with or without RAS flow)
- Effluent Temperature

This suggests that selectors can be successful even if they do not operate within the recommended ranges for some or even most of the parameters that are considered important for successful selector operation. This also suggests that MLSS, selector volume to total basin volume ratio, and aeration basin DO are important parameters, and operating within recommended ranges for these parameters may have been why EBMUD's selector was successful, while OCSD's selector system, which did not operate within the recommended ranges, was not.

Higher MLSS concentrations usually provide a higher concentration of bacteria in the selector, which may result in higher raCOD uptake rates. If this is true, then the selector HRT may not need to be as long. The higher uptake rate may also allow the selector MCRT to be shorter and may allow the selector to take a higher BOD_5/TSS influent without leakage. For any given F/M ratio, the concentration of both MLSS and BOD_5 can be very different (for the same basin volume and flow rate, an F/M = 1.0 could have a BOD_5 = 100 mg/L and MLSS = 100 mg/L, or a BOD_5 = 3,000 mg/L and MLSS = 3,000 mg/L. In the first example the concentrations of both BOD_5 and MLSS are much lower than in the second example. The second example with the higher BOD_5 and MLSS concentrations may drive much higher raCOD uptake rates and require less reaction time than the first example with much lower BOD_5 and MLSS concentrations). This may change the actual optimum F/M. If true, then keeping the MLSS concentration high may be very important to the success of a selector and would agree with the very high R^2 value obtained in the regression analysis.

Optimizing the selector volume/total basin volume ratio could promote optimum conditions for PAOs, and higher aeration basin DO can reduce the propensity for an activated sludge to grow filamentous organisms. It may also promote optimum conditions for PAOs. Nonetheless, these parameters may not have been as important if the selectors were operated within recommended ranges for reactor MCRT, selector F/M, or some of the other parameters considered. For example, the OCSD selector basins' SVI improved when the reactor MCRT was increased, even though the aeration basin DO decreased. Average MLSS also increased, and both selector and reactor F/Ms declined, becoming more compliant with recommended ranges.

Although this result may complicate selector design and operation, it shows that selector systems may be successful without operating within the recommended ranges for all the selector design and operating parameters. To simplify this analysis, a computerized selector diagnostic tool was prepared by the project team and is available on CD-ROM, which can be found on the inside back cover of this report. Documentation for this software can be found in Appendix F.

5.5.1 Conclusions

The comparison between the EBMUD and OCSD selector systems' operating values and recommended parameter ranges, determined in this study's regression analysis, resulted in the following conclusions:

- Selector installations can be successful even if operated outside of some or most of the recommended design/operating parameter ranges for successful selector operation.
- Selector installations will probably not be successful if operated outside of all the recommended design/operating parameter ranges for successful selector operation.
- Average MLSS appears to be an important parameter to keep within the recommended operating range. In the EBMUD/OCSD case, the selector volume/total basin volume ratio and aeration basin DO also appeared to be important parameters.
- Using the recommended parameter ranges for successful selector operation as a guide appears to offer good assistance to those who wish to determine why a selector is not performing as expected or to optimize a selector design.

The computerized selector diagnostic tool prepared for this project, and accessible through the CD-ROM on the back cover of this report, is an easy way to use this method for assistance in troubleshooting or designing a selector installation. Documentation for this software can be found in Appendix F of this report.

CHAPTER 6.0

SUMMARY AND CONCLUSIONS

6.1 Summary

Although the literature provides separate design and operating parameters for aerobic, anoxic, and anaerobic selectors, these parameters are assumed to be the same for either long- or short-MCRT activated sludge plants. Since distinctly different groups of filamentous bacteria predominate at short- versus long-MCRT—due in large part to differences in growth requirements between these two filament groups—they may require different control parameters. Consequently, the full-scale activated sludge plant data collected during this study were separated into long- and short-MCRT groups, based primarily on the type of filamentous bacteria present.

The short-MCRT plants were further split into two groups—plants equipped with aerobic selectors and plants equipped with either anoxic or anaerobic selectors—based on the hypothesis that aerobic selectors were more kinetically favorable to floc-forming bacteria than filamentous bacteria, and anaerobic or anoxic selectors were more metabolically favorable to floc-forming bacteria. The results of the regression analysis supported these differences among the three different WWTP groups—short-MCRT with anoxic or anaerobic selectors, short-MCRT with aerobic selectors, and long-MCRT. In general, selectors in the long-MCRT plants did not appear to reduce filamentous bulking (DSVI); in fact, the results suggest that unaerated selectors may enhance filamentous bulking in long-MCRT plants.

Selector design and operating parameters were quantitatively ranked according to their influence on DSVI, using the regression R^2 value. Using cubic polynomial regression curves, parameter ranges that were associated with the lowest (and highest) DSVI were determined. Many of these parameter ranges agreed well with those found in the literature. Some did not, for reasons that could be explained logically. Some design and operating parameters thought to be significant at the start of this study were instead found to have little if any influence on DSVI.

6.1.1 Long-MCRT Plants

MLSS

The most significant operating parameter for achieving low DSVIs in long-MCRT plants was average MLSS concentration—the higher the MLSS concentration, the lower the DSVI. This relationship held for short-MCRT plants with anaerobic and anoxic selectors. The influence of MLSS on DSVI could have been the result of a direct relationship through the DSVI calculation or an indirect relationship with another parameter that has a significant influence on DSVI, but MLSS had much less influence on DSVI in short-MCRT plants with aerobic selectors. Furthermore, higher MLSS concentrations in these plants corresponded to higher DSVIs, in contrast to the other WWTP groups.

Aerobic selectors rely on rapid uptake of raCOD[1] to select for floc-forming bacteria over filamentous bacteria. A very high oxygen consumption rate is associated with this high raCOD

[1] Refer to raCOD discussion on Page 1-3 in Chapter 1.0.

uptake. Oxygen is therefore at a premium in aerobic selectors, and lower MLSS will take up less oxygen than higher MLSS concentrations. Since this mechanism is not important in the other two WWTP groups, higher MLSS concentrations could be beneficial in supplying a higher concentration of floc-forming bacteria in the selector basin where substrate is added. Regardless of the mechanism, however, higher MLSS concentration is recommended for reducing DSVI in long-MCRT WWTPs, based on the regression analysis results from this study.

Selector Effect

The regression analysis showed that some parameters that have been considered vital to inducing a selector effect for filamentous bacteria control were not associated with lower DSVIs in long-MCRT activated sludge systems. Further, the regression analyses suggested that adding anoxic selectors to a fully-aerated basin would increase rather than decrease DSVI. The lowest DSVIs were associated with plants that approached a selector HRT = 0, an ICZ HRT = 0, a selector volume to total basin volume = 0, and the number of selector stages = 0. The selector F/M, selector MCRT, and the ICZ F/M all had an insignificant influence on DSVI in long-MCRT WWTPs. These results support the previous conclusion that selectors in long-MCRT WWTPs do not reduce DSVI, and anoxic selectors may promote higher DSVIs in these plants.

While this analysis suggests that the solution to DSVI problems in long-MCRT plants would be to remove any anoxic (and probably any anaerobic) selectors, unfortunately, most, if not all, of the plants in this group rely on them for nutrient removal.

Number of Aeration Basin Stages

DSVI decreased as the number of aeration basin stages (not selector stages) increased up to eight stages in these long-MCRT WWTPs.

Effluent pH and Temperature

The literature does not mention either pH or temperature as controlling factors for DSVI, except that filamentous fungus can grow in activated sludge at low pH. The regression analysis suggested that both pH and temperature can play a significant role in determining DSVI. The lowest DSVIs occurred at pHs of 6.4–6.7 and temperatures of 27°–32°C. The highest DSVIs were associated with pH values ≥7.7 and temperatures of 13°–17°C. Interestingly, pH values and temperatures associated with low DSVIs in long-MCRT systems were also associated with unfavorable growth conditions for *M. parvicella*, while temperatures and pH values associated with high DSVIs were also associated with favorable growth conditions for *M. parvicella*.

Long-MCRT Plant Summary

The regression analysis suggested that for the lowest DSVI, the best long-MCRT designed and operated plants had high MLSS (2,500–4,500$^+$ mg/L), compartmentalized aeration basins, and no anoxic or anaerobic zones (if nutrient removal was not needed). Further, DSVIs were lower when the pH = 6.4–6.7 and temperatures = 27°–32°C, and higher when pH = 7.7$^+$ and temperature = 13°–17°C.

6.1.2 Short-MCRT Plants with Anoxic or Anaerobic Selectors

Anoxic and anaerobic selectors rely on metabolic (rather than kinetic) competition for raCOD. Anoxic selectors remove raCOD through denitrification, while anaerobic selectors remove raCOD by using energy stored during a previous aerobic period. Aerobic selectors rely on rapid raCOD uptake induced by higher raCOD concentrations in aerobic selectors.

Selector Effect

The lowest DSVIs in short-MCRT WWTPs with anoxic or anaerobic selectors were associated with the lowest selector F/Ms, selector MCRT of 2–3 d or higher, the longest selector HRTs, two selector stages, an ICZ F/M as low as possible, and an ICZ HRT equal to the selector HRT. The regression results suggest that anaerobic or anoxic selectors should be as large as possible. The selector volume to total basin volume ratio results, however, limits the ideal selector size to 22.5%–25.0% of the total basin volume. The regression results suggest that the higher the MLSS concentration the lower the DSVI. The higher MLSS could enhance the effectiveness of the anaerobic or anoxic selector by providing a higher concentration of denitrifying or phosphorus removal bacteria for raCOD uptake.

Short-MCRT Plants with Anoxic or Anaerobic Selectors Summary

According to the regression analysis (see Table 4-10), the best design and operation of a short-MCRT activated sludge plant with anoxic or anaerobic selectors would include:

- a selector volume as large as possible while keeping the selector volume to total basin volume ratio between 22.5%–25.0%,
- two selector stages,
- a selector MCRT >2–3^{+} d,
- a MLSS concentration of 1,500–2,000^{+} mg/L,
- an aeration basin DO concentration between 2.5 and 4.0 mg/L, and
- as long a reactor MCRT as possible.

Other factors influencing DSVI include activated sludge influent BOD_5/TSS ratio (best is <0.5) and effluent temperature (best is 20°–25°C and worst is 27°–30°C, which matches well with the Type 1701 growth rate being higher than floc-forming bacteria at temperatures around 28°C and frequently less than floc-formers at temperatures less than 28°C, per Wanner, 1994).

6.1.3 Short-MCRT Plants with Aerobic Selectors

Selector Effect

Aerobic selectors function by having a high enough raCOD concentration to induce a kinetic selection of floc-forming bacteria over filamentous bacteria, so it is essential that the ICZ be small. The regression analysis results support this since very short ICZ HRTs (4.5–7.5 min calculated with RAS flows and 6.3–6.6 min calculated without RAS flows) are associated with lower DSVIs.

Recommended ranges for the ICZ F/M are included in the literature as well as total selector F/M and MCRT. The regression analysis, however, ranked these parameters as much less important as the ICZ HRT, and the literature has offered no recommended values to achieve lower DSVIs.

Other Parameters

The activated sludge influent BOD_5 concentration was the most influential parameter on DSVI, with high influent BOD_5 concentrations being associated with high DSVIs. This association may exist because with higher influent BOD concentrations, there is a greater likelihood that raCOD will leak into the main aeration basin and cause filamentous bulking. The lowest DSVIs were associated with influent BOD_5 values <80 mg/L. Low effluent pH (6.3–6.6) was also associated with lower DSVI; however, no recommendation to modify pH is implied. The RAS flow rate associated with the lowest DSVIs was 25°–35% of influent flow. Higher RAS flow rates may dilute the raCOD and reduce the kinetic benefits of an aerobic selector. Contrary to the findings for anoxic and anaerobic selector plants, low reactor MCRTs were associated with the lowest DSVIs.

Short-MCRT Plants with Aerobic Selectors Summary

The ICZ must be small enough to provide a high enough raCOD to induce kinetic selection of floc-forming bacteria over filamentous bacteria. Although higher influent BOD_5 concentrations may result in raCOD bleeding through a selector, the ICZ F/M does not appear to be the most important design and operating parameter for a successful aerobic selector. Further, the %RAS should be as low as possible (25%–35%), the MLSS should be as low as possible (to about 1,000 mg/L), the reactor MCRT should be low (<1.3 d), and the aeration basin DO should be high.

6.2 Conclusions

Laboratory Investigation

- Severe bulking (DSVI ≥500 mL/g) due to *Thiothrix* spp. was controlled by installing three-stage and four-stage aerobic selectors or by removing raCOD from the wastewater fed to an activated sludge process. This suggests that removing raCOD prior to the main activated sludge aeration basin, either with a selector or by excluding it from a synthetic sewage fed to the activated sludge process, will significantly reduce the growth of *Thiothrix* spp.
- Removing raCOD from wastewater fed to activated sludge processes alone may not produce DSVIs as low as activated sludge processes equipped with a well-performing selector. This may be because selectors enhance the growth of raCOD floc-forming bacteria, while activated sludge processes fed wastewaters absent of raCOD do not support the growth of these organisms; and raCOD floc-forming bacteria may enhance activated sludge floc structure and settling on their own.
- Uptake rates for Tween 80, and possibly LCFAs, were six–10 times slower than uptake rates for acetate. This suggests that even well-performing selectors may not adequately remove LCFAs and could allow them to leak into the main aeration basin where they may be used by filamentous organisms for growth.

Detailed Plant Investigations

Comparing average selector design and operating data versus 90th percentile SVI and DSVI yielded the following conclusions:

- Anoxic selector installations demonstrated superior settleability control compared to anaerobic selectors. Approximately 85% of anoxic selector plants (23 of 27) had 90th percentile DSVIs <150 mL/g, while only 14% of anaerobic plants (two of 14) achieved this result. Most anoxic selectors, however, were installed in long-MCRT plants, while all anaerobic selectors were installed in short-MCRT plants. The lower DSVI in plants with anoxic selectors may be because of MCRT and the type of filamentous bacteria that grow at long MCRT (Wanner, 1994) rather than selector type.

- Selector staging was not observed to have a significant impact on settleability in anoxic selector systems. In fact, all eight single-stage anoxic selectors yielded 90th percentile DSVIs <150 mL/g, while four of 18 multi-stage anoxic selectors exceeded this value.
- Selector staging was not observed to have a significant impact on settleability in anaerobic selector systems, since six of seven systems yielded 90th percentile DSVIs >150 mL/g in both the single- and multi-stage categories.
- Based on plots of average values vs. 90th percentile DSVIs, no significant relationships were identified between settleability and selector ICZ F/M, selector F/M, selector MCRT, reactor MCRT (excluding secondary clarifier solids), contact loading, or selector HRT.

Comparing average parameter and 90th percentile SVI/DSVI values for the plants included in the detailed plant investigation is somewhat limited since each facility is represented by only a single data point that does not reflect variation in each parameter. A single-variable regression analysis, incorporating daily operating data for each facility, was conducted to better evaluate the influence of parameter variation on SVI and DSVI values.

The regression analysis for short-MCRT WWTPs with anoxic or anaerobic selectors, short-MCRT WWTPs with aerobic selectors, and long-MCRT WWTPs yielded the following main conclusions:

- For short-MCRT WWTPs, anaerobic and anoxic selectors should be sized large enough to remove all or most of the raCOD and should be staged to prevent short-circuiting and raCOD breakthrough to the main aeration basin (rather than to provide a kinetic advantage).
- For short-MCRT WWTPs with aerobic selectors, an raCOD concentration gradient is required to provide a kinetic advantage to floc-formers over filamentous organisms; however, at higher influent BOD_5 concentrations, sufficient raCOD may leak through to the main aeration zone to cause bulking problems.
- Selectors do not significantly control filamentous organisms and bulking in long-MCRT plants, which is supported in the literature (Wanner, 1993; Jenkins et al., 2004; Martins et al., 2004b).

Full-Scale Demonstration Projects

The comparison between the EBMUD and OCSD selector systems' operating values and recommended parameter ranges, determined in this study's regression analysis, yielded the following conclusions:

- Selector installations can be successful even if operated outside of some or most of the recommended design/operating parameter ranges for successful selector operation.
- Selector installations will probably not be successful if operated outside of all the recommended design/operating parameter ranges for successful selector operation.
- Average MLSS appears to be an important parameter to keep within the recommended operating range. In the EBMUD/OCSD case, the selector volume/total basin volume ratio and aeration basin DO also appeared to be important parameters.

- Using the recommended parameter ranges for successful selector operation as a guide appears to offer good assistance to those who wish to determine why a selector is not performing as expected or to optimize a selector design.

Computerized Selector Diagnostic Tool

The computerized selector diagnostic tool prepared for this project, and accessible through the CD-ROM on the back cover of this report, is an easy way to use this method for assistance in troubleshooting or designing a selector installation. Documentation for this software can be found in Appendix F of this report.

APPENDIX A

INITIAL SCREENING AND DETAIL PLANT INVESTIGATION SURVEY FORMS

WATER ENVIRONMENT RESEARCH FOUNDATION

"Develop and Demonstrate Fundamental Basis for Selectors to Improve Activated Sludge Settleability" (01-CTS-4)

INITIAL SCREENING SURVEY

PLANT NAME:

LOCATION:

CONTACT NAME:

PHONE:

E-MAIL:

GENERAL PLANT INFORMATION

Flow (MGD): Industrial Contribution: ☐ < 15% of Total BOD ☐ ≥ 15% of Total BOD

Major Industrial Contributor:

Yearly Temperature Variation (°C) Low: High:

Nutrient Removal (check all that apply): ☐ NH_3 ☐ NO_3 ☐ P

Biological or Chemical:

ACTIVATED SLUDGE SYSTEM CONFIGURATION

Influent Feed (check box): ☐ Raw ☐ Primary Effluent

Operating Solids Residence Time (days):

Aeration Basin Configuration: (check box)

☐ Complete Mix

☐ Plug Flow with Compartments → No. of Compartments

☐ Plug Flow without Compartments → Approximate L:W Ratio

Description of approximate dissolved oxygen profile through aeration basin:

Selector Present? ☐ Yes ☐ No If yes, what type?[a] ☐ Aerobic ☐ Anoxic ☐ Anaerobic *(see footnote for definition of terms)*

ACTIVATED SLUDGE SETTLEABILITY

Problems with sludge settleability? ☐ Yes ☐ No

If yes, type of problem: ☐ Filamentous Bulking ☐ Viscous Bulking ☐ Other

Bulking Frequency: ☐ < 1% ☐ < 5% ☐ 5-20% ☐ 20-35% ☐ 35-50% ☐ 50-75% ☐ >75%

Microscopic ID of responsible organisms? ☐ Yes ☐ No

Type identified:

Approximate SVI (mL/g): Typical During bulking incident

If present, did selector help improve sludge settleability? ☐ Yes ☐ No

Additional Comments:

---

[a] Aerobic: Selector zone is aerated with measurable dissolved oxygen concentration
Anoxic: Selector zone is unaerated, but has nitrate present (either from RAS or internal mixed liquor recycle)
Anaerobic: Selector zone is unaerated and no nitrate is present

WERF Project 01-CTS-4 - "Develop and Demonstrate Fundamental Basis for Selectors to Improve Activated Sludge Settleability"

DETAILED PLANT SURVEY – AEROBIC SELECTORS

PLANT NAME: ______________________ CONTACT: ______________________

LOCATION: ______________________ PHONE: ______________________

E-MAIL: ______________________

SECONDARY PROCESS LAYOUT (see attached sketch for description of terms)

- Upstream Biological Processes ______________________
- No. of Aeration Basins ______________
  - Peak Flow per Basin ______________
  - Avg. Flow per Basin ______________
- No. of Selector Stages ______________
- Size of Selector Stages ______________ Stage 1
  (L x W x water depth in ft) ______________ Stage 2
  ______________ Stage 3
  ______________ Stage 4
- No. of Remaining Stages ______________
- Size of Remaining Stages ______________
  (L x W x water depth in ft)
- Total Selector Zone Volume (MG) ______________
- Total Main Aeration Zone Volume (MG) ______________
- Type of Mixing/Aeration in Selector ______________

Return Flows

- Internal Recycle Streams? ☐ Yes ☐ No
  - Type ______________
- RAS Feed Location?
  - Upstream of Selector ☐
  - At Selector ☐
- Multiple RAS Feed Points? ☐ Yes ☐ No
  - Describe: ______________

Type of Aeration System

☐ Diffused Air (fine bubble)
☐ Diffused Air (coarse bubble)
☐ Mechanical Aerators
- Horsepower (hp) Profile

____

☐ Other ______________

PROCESS DATA COLLECTION

- One year's worth of daily operational data on Secondary Influent, including:
  - BOD
  - Soluble BOD
  - COD
  - Soluble COD
  - TKN
  - P
- One year's worth of daily operational data for the following key parameters:
  - Secondary Influent Flow
  - No. of Basins In-Service
  - RAS Flow
  - RAS Conc.
  - MLSS
  - MLVSS
  - Total System SRT
  - Aerated SRT
  - WAS Flow
  - WAS Conc.
  - SVI or DSVI
  - F/M
  - DO
  - Inf./Eff. pH
  - Filament Type/Abundance

ADDITIONAL PLANT INFORMATION

- Please provide the following information, if available:
  - Process Schematic
  - Selector Design Parameters
  - Selector Evaluation Data/Reports
  - Secondary Treatment O&M Manual (excerpts)
  - DO Profile Data across Aeration Basin
  - Secondary Influent H_2S Conc. (typ.)
  - Soluble BOD or COD Exiting Selector Zone
  - Oxygen Uptake Rate Data

WERF Project 01-CTS-4 - "Develop and Demonstrate Fundamental Basis for Selectors to Improve Activated Sludge Settleability"

DETAILED PLANT SURVEY – ANOXIC SELECTORS

PLANT NAME: ______________________ CONTACT: ______________________

LOCATION: ______________________ PHONE: ______________________

E-MAIL: ______________________

SECONDARY PROCESS LAYOUT (see attached sketch for description of terms)

- Upstream Biological Processes ______________
- No. of Aeration Basins ______________
  - Peak Flow per Basin ______________
  - Avg. Flow per Basin ______________
- No. of Selector Stages ______________
- Size of Selector Stages ______________ Stage 1
  (L x W x water depth in ft) ______________ Stage 2
  ______________ Stage 3
  ______________ Stage 4
- No. of Remaining Stages ______________
- Size of Remaining Stages ______________
  (L x W x water depth in ft)
- Total Selector Zone Volume (MG) ______________
- Total Main Aeration Zone Volume (MG) ______________
- Type of Mixing in Selector Zone ______________

Return Flows

- Internal Recycle Streams? ☐ Yes ☐ No
  - Type ______________
- RAS Feed Location?
  - Upstream of Selector ☐
  - At Selector ☐
- Multiple RAS Feed Points? ☐ Yes ☐ No
  - Describe: ______________

Type of Aeration System

☐ Diffused Air (fine bubble)
☐ Diffused Air (coarse bubble)
☐ Mechanical Aerators
- Horsepower (hp) Profile ______________

☐ Other ______________

PROCESS DATA COLLECTION

- One year's worth of daily operational data on Secondary Influent, including:
  - BOD
  - Soluble BOD
  - COD
  - Soluble COD
  - TKN
  - P
- One year's worth of daily operational data for the following key parameters:
  - Secondary Influent Flow
  - No. of Basins In-Service
  - RAS Flow
  - RAS Conc.
  - Filament Type/Abundance
  - MLSS
  - MLVSS
  - Total System SRT
  - Aerated SRT
  - Nitrate Recycle Flow
  - WAS Flow
  - WAS Conc.
  - SVI or DSVI
  - F/M
  - DO
  - Effluent NO_3
  - Effluent PO_4
  - Inf./Eff. pH

ADDITIONAL PLANT INFORMATION

- Please provide the following information, if available:
  - Process Schematic
  - Selector Design Parameters
  - Selector Evaluation Data/Reports
  - Secondary Treatment O&M Manual (excerpts)
  - DO Profile Data across Aeration Basin
  - Secondary Influent H_2S Conc. (typ.)
  - Soluble BOD or COD Exiting Selector Zone
  - Oxygen Uptake Rate Data

WERF Project 01-CTS-4 - "Develop and Demonstrate Fundamental Basis for Selectors to Improve Activated Sludge Settleability"

DETAILED PLANT SURVEY – ANAEROBIC SELECTORS

PLANT NAME: ______________________ CONTACT: ______________

LOCATION: ______________________ PHONE: ______________

E-MAIL: ______________

SECONDARY PROCESS LAYOUT (see attached sketch for description of terms)

- Upstream Biological Processes ______________
- No. of Aeration Basins ______________
  - Peak Flow per Basin ______________
  - Avg. Flow per Basin ______________
- No. of Selector Stages ______________
- Size of Selector Stages ______________ Stage 1
  (L x W x water depth in ft) ______________ Stage 2
  ______________ Stage 3
  ______________ Stage 4
- No. of Remaining Stages ______________
- Size of Remaining Stages ______________
  (L x W x water depth in ft)
- Total Selector Zone Volume (MG) ______________
- Total Main Aeration Zone Volume (MG) ______________
- Type of Mixing in Selector Zone ______________

Return Flows

- Internal Recycle Streams? ☐ Yes ☐ No
  - Type ______________
- RAS Feed Location?
  - Upstream of Selector ☐
  - At Selector ☐
- Multiple RAS Feed Points? ☐ Yes ☐ No
  - Describe: ______________

Type of Aeration System

☐ Diffused Air (fine bubble)

☐ Diffused Air (coarse bubble)

☐ Mechanical Aerators

- Horsepower (hp) Profile

____

☐ Other ______________

PROCESS DATA COLLECTION

- One year's worth of daily operational data on Secondary Influent, including:
  - BOD
  - Soluble BOD
  - COD
  - Soluble COD
  - TKN
  - P
- One year's worth of daily operational data for the following key parameters:
  - Secondary Influent Flow
  - No. of Basins In-Service
  - RAS Flow
  - RAS Conc.
  - MLSS
  - MLVSS
  - Total System SRT
  - Aerated SRT
  - WAS Flow
  - WAS Conc.
  - SVI or DSVI
  - F/M
  - DO
  - Effluent PO_4
  - Inf./Eff. pH
  - Filament Type/Abundance

ADDITIONAL PLANT INFORMATION

- Please provide the following information, if available:
  - Process Schematic
  - Selector Design Parameters
  - Selector Evaluation Data/Reports
  - Secondary Treatment O&M Manual (excerpts)
  - DO Profile Data across Aeration Basin
  - Secondary Influent H_2S Conc. (typ.)
  - Soluble BOD or COD Exiting Selector Zone
  - Oxygen Uptake Rate Data

WERF Project 01-CTS-4 - "Develop and Demonstrate Fundamental Basis for Selectors to Improve Activated Sludge Settleability"

DETAILED PLANT SURVEY

EXPLANATION OF SELECTOR CONFIGURATION TERMINOLOGY

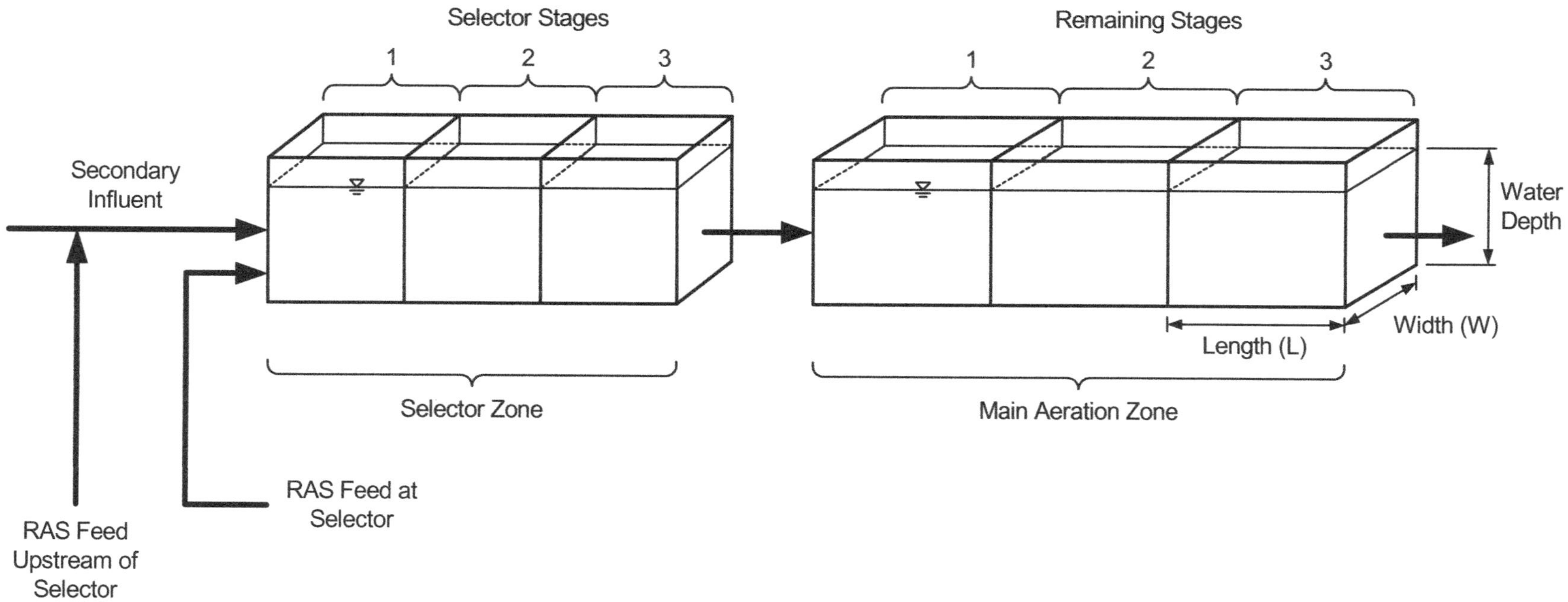

APPENDIX B

SUMMARY OF OPERATING CONDITIONS IDENTIFIED WITH COMMON FILAMENTOUS ORGANISMS

This section summarizes the characteristics of important filamentous bacteria, and the operating conditions associated with their presence in activated sludge treatment (per the literature review presented in Chapter 2.0). Each filamentous organism type is grouped according to whether it is controlled with a selector, not controlled with a selector, or unknown if it is controlled with a selector.

FILAMENTOUS ORGANISMS REPORTEDLY CONTROLLED WITH SELECTORS

Sphaerotilus natans are obligate aerobes and rarely, if ever, occur in biological nutrient removal (BNR) plants. However, it regularly occurs in industrial WWTPs and grows well in bench- and pilot-scale systems. This species is favored by low DO conditions or nutrient deficiency (N or P). Readily assimilable COD, high sludge loading levels [>0.2 kg BOD_5/(kg MLSS·d)], short MCRT, and higher temperature favor *S. natans* growth.

Type 021N is an obligate aerobe, frequently causing bulking sludge in domestic and industrial plants with nutrient removal. It can grow well at a broad range of sludge loading levels [0.05–0.4 kg BOD_5/(kg MLSS·d)]. Readily assimilable substrates, nutrient deficiency, and low DO concentration favor Type 021N growth.

Type 1701 was one of the most commonly observed filaments in the 1980s in the U.S. It rarely, if ever, occurs in BNR treatment plants treating a domestic wastewater. Moderate sludge loading levels [>0.2 kg BOD_5/(kg MLSS·d)], short MCRT, a high level of carbohydrates in wastewater, low DO concentration, and relatively high temperatures (>15°C) favor Type 1701 growth.

Thiothrix spp. usually occurs in moderately loaded domestic plants [>0.17 kg BOD_5/(kg MLSS·d)] and industrial wastewater treatment plants. Besides consuming raCOD, this filament also uses reduced sulfur compounds for growth. Nutrient deficiency (N and P) and low DO concentration are favorable to this species. Nielsen et al. (2000) studied the metabolism of *Thiothrix* in mixed liquor from an industrial wastewater treatment plant with severe bulking problems. The plant received easily degradable wastewater with a high content of low molecular weight alcohols, organic acids, and other compounds from food additives. It contained some sulfides, but the level was not quantified. The activated sludge system was operated at an SRT of 8–10 d and a temperature of 15–25°C. The DO concentration ranged from 0.5 to 2.0 mg/L. Some ammonia was added intermittently to provide nitrogen to the activated sludge. By using a combination of fluorescence in situ hybridization and microautoradiography, Nielsen et al. were able to study the in situ metabolism of the *Thiothrix* filaments under different environmental conditions. The organism was very versatile and was able to incorporate acetate and bicarbonate under heterotrophic mixotrophic metabolism under anaerobic conditions with or without nitrate. Intracellular sulfur globules were formed from thiosulfate and acetate uptake. The doubling time of the organism was 6–9 h under mixotrophic conditions and 15–30 h under autotrophic conditions. A key property that provided a benefit for *Thiothrix* growth was the ability to take up acetate in the presence of thiosulfate.

Type 0961 is not frequently observed in activated sludge, but it sometimes causes bulking sludge in industrial wastewater treatment plants. Moderate sludge loading levels [>0.2 kg BOD_5/(kg MLSS·d)], long MCRTs (>6 d), raCOD, and high temperatures favor Type 0961.

Haliscomenobacter hydrossis has a small role in domestic treatment plants, but grows well in industrial treatment plants. Low DO concentration and phosphorus deficiency favors this filamentous organism. Moderate loading levels [>0.2 kg BOD_5/(kg MLSS·d)] for domestic wastewater, lower loading levels for industrial wastewater, high influent nitrogen, and raCOD favor *H. hydrossis* growth. This organism was found to grow rapidly with increased SVI in a completely-mixed activated sludge (CMAS) system with a high transient substrate overload and an oxygen deficit (Pernelle et al., 2001).

FILAMENTOUS ORGANISMS NOT LIKELY CONTROLLED WITH SELECTORS

Microthrix parvicella is the most frequent cause of bulking sludge in many countries. The species commonly occurs in lowly loaded [<0.2 kg BOD_5/(kg MLSS·d)] and long-MCRT domestic treatment plants. Low DO concentration, LCFAs (e.g. oleic acid), low temperatures, high influent reduced sulfur compounds, and high influent nitrogen compounds can favor this species. It seldom occurs in industrial treatment plants.

Type 0041 occurs very commonly in activated sludge. Nutrient deficiency, low sludge loading levels [<0.2 kg BOD_5/(kg MLSS·d)], long MCRT, saCOD, and the absence of primary treatment are favorable to Type 0041.

Type 0092 has been associated with the presence of *M. parvicella*. Type 0092 was the most common filamentous organism in South African activated sludges, especially BNR activated sludges (Blackbeard, et al., 1986). Similar to *M. parvicella*, it prefers a low F/M operating condition [< 0.1 kg BOD_5/(kg MLSS·d)] and long MCRT. However, Type 0092 prefers a warmer temperature (>15°C), and may require anoxic and anaerobic zones. In many treatment plants, the disappearance of *M. parvicella* in late spring is coupled with an increase in Type 0092 filaments (Eikelboom, 2000). Krhutkova et al. (2002) noted a significant occurrence of this filament in a survey of eight Czech Republic plants in 2000 and generally observed its presence with *M. parvicella*.

Type 0675 is a common filamentous organism in activated sludge treatment plants in many countries. It is nearly always found in activated sludge plants treating wastewater from pulp and paper plants. Type 0675 is often observed with Type 0041. Nutrient deficiency, long MCRT, and saCOD are favorable to Type 0675.

Type 1851 is regularly observed in lowly-loaded plants and was the sixth most frequently observed dominant filament in BNR plants in South Africa (Blackbeard et al., 1987). The species can grow well in industrial treatment plants (agricultural industry). Low sludge loading levels [<0.15 kg BOD_5/(kg MLSS·d)], long MCRT, raCOD (especially simple sugars), and relatively high temperature (25–30°C) are favorable to Type 1851. Some have reported that Type 1851 is controlled with a selector.

Type 0914 is observed in South African and Danish activated sludge plants. High influent reduced sulfur, low sludge loading levels [<0.2 kg BOD_5/(kg MLSS·d)], and long MCRT are favorable to Type 0914. Type 0914 may also grow at higher temperatures (up to 50°C).

Nocardia spp. is the most common filamentous organism in U.S. activated sludge, notorious for forming scum, but is not responsible for sludge settleability problems. Surface active materials, fat, and internal recycling of any floating material causes a formation of foam and scum. Relatively high temperatures (>15°C) and moderate sludge loading levels [0.1–0.7 kg BOD_5/(kg MLSS·d)] can favor this species. Similar to *M. parvicella*, these organisms are able to grow on LCFAs (Chua et al., 1994, 1996). Using pure cultures, *Nocardia* growth yields of 0.12, 0.16, 0.14, and 0.18 g VSS/g fatty acid were observed for nonanoic, undecanoic, palmitic, and stearic acids, respectively.

Nostocoida limicola spp. are frequently observed in activated sludge plants, especially at higher loadings and in industrial plants.. Readily assimilable substrates (especially simple sugars) and slowly assimilable substrates (long chain fatty acids) are both utilized, phosphorus deficiency, low temperatures, anoxic or anaerobic zones, moderate sludge loading levels [0.1–0.3 kg BOD_5/(kg MLSS·d)], and long MCRT are favorable to *N. limicola*.

UNKNOWN WHETHER SELECTORS ARE EFFECTIVE OR NOT

Type 0411 is occasionally observed in moderately-loaded treatment plants levels [0.2–0.7 kg BOD_5/(kg MLSS·d)] with high DO concentration conditions. This species only has a small effect on sludge settleability.

Type 1863 is principally observed at high sludge loading levels [>0.7 kg BOD_5/(kg MLSS·d)], and short MCRT. The species also contributes to scum formation, is present during startup of activated sludge plants, and disappears when the process becomes stabilized. Turbid effluents are often observed when the filament grows freely in suspension or surrounding the flocs. It is a strict aerobe, grows on VFAs as its sole carbon source (it cannot grow on carbohydrates), and produces polyphosphate inclusions. This organism usually has little impact to sludge settleability.

Type 0581 is occasionally observed in lowly-loaded domestic treatment plants with intermittent influent flow. The factors determining the growth of this filamentous species in treatment plants are not well known. It is often mistaken for *M. parvicella* under a wet mount slide.

Type 0211 is occasionally observed in highly-loaded activated sludge plants. The factors determining the growth of this filamentous species in treatment plants are not well understood.

Type 0803 is frequently present in domestic and industrial wastewater treatment plants. Low sludge loading levels [<0.2 kg BOD_5/(kg MLSS·d)], low DO concentrations, and anaerobic conditions are favorable to Type 0803.

APPENDIX C

DESCRIPTION OF PROCESS DATA CALCULATIONS FOR REGRESSION ANALYSIS DATA SETS

The following is a description of the data fields and process data calculations used primarily for the data sets in the regression analyses.

Field	Description
Flow (MGD)	Flow rate the through secondary system, not including RAS recycle or mixed-liquor recycle.
Flow+RAS (MGD)	Flow rate through the secondary system, including RAS recycle flow, but not including mixed-liquor recycle.
Plant Avg Flow (MGD)	The average flow through the secondary system during the study period. For each data set, this value is the same for the duration of the study period.
NormalizedFlow	[Flow (MGD)] / [Plant Avg Flow (MGD)]. The average should equal 1 throughout the study period for each data set.
AS Source	0 for preliminary treatment only, 1 for primary treatment. One plant (Yakima WWTP, #014) had primary treatment plus biological treatment (trickling filter) upstream of the activated sludge system, which was also categorized as 1.
AS Inf BOD_5 (mg/L) calc	Influent BOD_5 concentration to the activated sludge system. For systems without primary treatment, plant influent BOD_5 was used. For systems with primary treatment, primary effluent BOD_5 or secondary influent BOD_5 was used. For plants that use other organic loading measures instead of BOD_5 (e.g., COD or $cBOD_5$), the data was converted to BOD_5 units using the best available plant-specific data. In a limited number of cases, due to unavailability of data for conversion of secondary influent data to BOD_5 units, a conversion factor based on plant influent flow characteristics was applied to secondary influent parameters. "Calc" indicates that the data includes estimated values as substitutes for missing values, as described separately in Section 4.3.2.6.
AS Inf TSS (mg/L) calc	Influent TSS concentration to the activated sludge system. For systems without primary treatment, plant influent TSS was used as input. For systems with primary treatment, primary effluent TSS or secondary influent TSS was used as input. "Calc" indicates that the data includes estimated values as substitutes for missing values, as described separately in Section 4.3.2.6.

Field	Description
BOD_5/TSS Ratio	[AS Inf BOD_5 (mg/L) calc] / [AS Inf TSS (mg/L) calc]
BOD_5 Load (lb BOD_5/day)	[AS Inf BOD_5 (mg/L) calc] * [Flow (MGD)] * 8.34
Avg MLSS (mg/L) calc	Average MLSS concentration in the activated sludge system. Where data from multiple trains, or multiple compartments within a train was provided, these data were averaged. MLSS concentration along an aeration train was generally observed (and assumed) to be constant. Thus this average value was also used in calculations such as Sx1 F/M (associated with MLSS concentration in first zone) and Merkel DSVI (associated with MLSS concentration in the last zone). One exception was Davenport WPCP (#056), which is a contact stabilization plant in which the contact zone MLSS and the stabilization zone MLSS differ significantly. As with other plants, the Avg MLSS (mg/L) for Davenport was calculated (volume-weighted average), but was not used for calculations where the MLSS in a specific zone was required (such as for Sx1 F/M or DSVI calculations). "Calc" indicates that the data includes estimated values as substitutes for missing values, as described separately in Section 4.3.2.6.
SVI (mL/g)	Where SVI data was provided separate from each aeration train, the SVI values were averaged.
DSVI (mL/g) calc	DSVI calculated by Merkel equation (calculation described separately). Three plants provided measured DSVI data, which was used as is (no Merkel calculation). "Calc" indicates that the data includes estimated values as substitutes for missing values, as described separately in Section 4.3.2.6.
Temp (°C) calc	Wastewater temperature in degrees Celsius. Where temperature data was obtained from multiple locations, reactor temperature was used preferentially over effluent temperature (second choice) or influent temperature (third choice). "Calc" indicates that the data includes estimated values as substitutes for missing values, as described separately in Section 4.3.2.6.
Sx1 Vol in Svc (MG)	The total volume of Sx1 compartments in use on a particular day. Out-of-service compartments and/or trains are excluded.
Sx2 Vol in Svc (MG)	The total volume of Sx2 compartments in use on a particular day. Out-of-service compartments and/or trains are excluded. Default is Null if Sx2 does not exist.
Sx3 Vol in Svc (MG)	The total volume of Sx3 compartments in use on a particular day. Out-of-service compartments and/or trains are excluded. Default is Null if Sx3 does not exist.
Sx4 Vol in Svc (MG)	The total volume of Sx4 compartments in use on a particular day. Out-of-service compartments and/or trains are excluded. Default is Null if Sx4 does not exist.
Sx5 Vol in Svc (MG)	The total volume of Sx5 compartments in use on a particular day. Out-of-service compartments and/or trains are excluded. Default is Null if Sx5 does not exist.

Field	Description
Sx6 Vol in Svc (MG)	The total volume of Sx6 compartments in use on a particular day. Out-of-service compartments and/or trains are excluded. Default is Null if Sx6 does not exist.
No of Sltr Stages	Number of selector stages in series. Where this number differed from one train to another, the data was analyzed as two separate data sets (e.g., 127-1 and 127-2).
Tot Sltr Vol in Svc (MG)	The total volume of all selector compartments in use on a particular day. Out-of-service compartments and/or trains are excluded.
Efftv ICZ Vol in Svc (MG)	Effective values derived on a case by case basis as specified in Selector Dimension Data.xls. Value is never larger than the actual [Sx1 Vol in Svc (MG)].
Efftv ICZ F/M	The effective ICZ F/M (lb BOD_5/lb MLSS-d), calculated as [Flow (MGD)]*[AS Inf BOD_5 (mg/L) calc]/([Efftv ICZ Vol in Svc (MG)]*[Avg MLSS (mg/L) calc]. For the Davenport WPCP (#056), the contact zone MLSS was used instead of the average MLSS.
Efftv ICZ Nom HRT	The nominal HRT (h) of the effective ICZ area, excluding RAS flow and IR flow. Calculated as [Efftv ICZ Vol in Svc (MG)] /[Flow (MGD)]*24.
Efftv ICZ Est Real HRT	The estimated real HRT (h) of the effective ICZ area, including RAS flow and IR flow. Calculated as [Efftv ICZ Vol in Svc (MG)] /[Flow (MGD)]*24.
Efftv No of Sltr Stages	Effective values derived on a case by case basis as specified in Selector Dimension Data.xls. Value is never smaller than [No of Sltr Stages].
No of AB Stages	Number of main aeration basin in series. The main aeration basin is defined as the portion of the train that is not the selector, and is not based on use of aeration. (Therefore, unaerated compartments that are not part of the selector are also included.) Where this number varied or differed from train to train, an average value was used.
Tot AB Vol (MG)	The total volume of all main aeration basin compartments in use on a particular day. Out of service compartments and/or trains are excluded.
Extra Vol (MG)	Extra volume that should be added [Tot Sltr Vol in Svc (MG)] and [Tot AB Vol (MG)] to get [Tot Reac Vol (MG)], but does not qualify as selector volume or main aeration basin volume. Created 2/4/06 to account for Winston-Green's winter operational mode, in which the 1st 2 compartments are used as solids holding compartments.
Oxy Ditch	1 for presence of oxidation ditch in the aeration train, 0 for absence of an oxidation ditch.

Field	Description
IR line	1 indicates use of mixed liquor recycle line from an aeration basin compartment back to one or more selector compartments. 0 indicates that such a line is not present (or if present is was not used during the study period).
IR to Sx1	1 indicates use of mixed liquor recycle line from an aeration basin compartment back to Sx1 (at least partial flow). 0 indicates that such a line is not present (or if present is was not used during the study period).
No of Trains in Svc	Number of (parallel) trains in service.
Wastage (lb SS/d)	Pounds of solids wasted per day. Used to calculate MCRT values. 0 value means wasting did not occur or was very minimal. Blank cell indicates that wasting data was not available. A common situation was that wasting occurred daily, but WAS SS concentration was only measured certain days of the week. In this situation, WAS SS concentration was estimated on days on which it was not measured. The estimate of the WAS SS concentration was either the average for the entire study period (for situations where the WAS SS concentration remained relatively constant throughout the period), or was an average of surrounding values (for situations where the WAS SS concentration trended up and down during the study period).
7d_Wastage(lb SS/d)	An average of [Wastage (lb SS/d)] over the last 7 days, up to and including the current day. This value was used to calculate 7-day MCRT values. The first 6 days of each data set are necessarily blank.
RAS Flow (MGD)	RAS flow rate from clarifier back to the beginning of the aeration train. The majority of the plants (~90%) supplied daily data for this field, but estimated values were used where daily data could not be obtained.
RAS SS (mg/L)	RAS SS concentration. Where separate WAS SS and RAS SS concentrations were not provided, this value was also assumed to be equal to WAS SS.
Est IR Flow	Estimated internal mixed liquor recycle flow. Values are estimated b/c none of the plants were able to provide daily flow based on meter readings.
Sx1 % IR Q	Percent of internal recycle flow flowing through Sx1.
Sx2 % IR Q	Percent of internal recycle flow flowing through Sx2. Value is cumulative and includes flow introduced in upstream selector compartments.
Sx3 % IR Q	Percent of internal recycle flow flowing through Sx3. Value is cumulative and includes flow introduced in upstream selector compartments.
Sx4 % IR Q	Percent of internal recycle flow flowing through Sx4. Value is cumulative and includes flow introduced in upstream selector compartments.

Field	Description
Sx5 % IR Q	Percent of internal recycle flow flowing through Sx5. Value is cumulative and includes flow introduced in upstream selector compartments.
Sx6 % IR Q	Percent of internal recycle flow flowing through Sx6. Value is cumulative and includes flow introduced in upstream selector compartments.
Sx1 Est Tot Q	Estimated total flow through Sx1, calculated as [Flow (MGD)] + [RAS Flow (MGD)] + ([Est IR Flow]*[Sx1 % IR Q]/100). Units in MGD.
Sx2 Est Tot Q	Estimated total flow through Sx2, calculated as [Flow (MGD)] + [RAS Flow (MGD)] + ([Est IR Flow]*[Sx2 % IR Q]/100). Units in MGD.
Sx3 Est Tot Q	Estimated total flow through Sx3, calculated as [Flow (MGD)] + [RAS Flow (MGD)] + ([Est IR Flow]*[Sx4 % IR Q]/100). Units in MGD.
Sx4 Est Tot Q	Estimated total flow through Sx4, calculated as [Flow (MGD)] + [RAS Flow (MGD)] + ([Est IR Flow]*[Sx4 % IR Q]/100). Units in MGD.
Sx5 Est Tot Q	Estimated total flow through Sx5, calculated as [Flow (MGD)] + [RAS Flow (MGD)] + ([Est IR Flow]*[Sx5 % IR Q]/100). Units in MGD.
Sx6 Est Tot Q	Estimated total flow through Sx6, calculated as [Flow (MGD)] + [RAS Flow (MGD)] + ([Est IR Flow]*[Sx6 % IR Q]/100). Units in MGD.
Sx1 Est Real HRT	Estimated real HRT (h) thru Sx1, calculated as [Sx1 Vol in Svc (MG)] / [Sx1 Est Tot Q] *24.
Sx2 Est Real HRT	Estimated real HRT (h) thru Sx2, calculated as [Sx2 Vol in Svc (MG)] / [Sx2 Est Tot Q] *24.
Sx3 Est Real HRT	Estimated real HRT (h) thru Sx3, calculated as [Sx3 Vol in Svc (MG)] / [Sx3 Est Tot Q] *24.
Sx4 Est Real HRT	Estimated real HRT (h) thru Sx4, calculated as [Sx4 Vol in Svc (MG)] / [Sx4 Est Tot Q] *24.
Sx5 Est Real HRT	Estimated real HRT (h) thru Sx5, calculated as [Sx5 Vol in Svc (MG)] / [Sx5 Est Tot Q] *24.
Sx6 Est Real HRT	Estimated real HRT (h) thru Sx6, calculated as [Sx6 Vol in Svc (MG)] / [Sx6 Est Tot Q] *24.
Tot Sltr Est Real HRT	Estimated real HRT (h) thru the entire selector, calculated as the sum of [Sx1 Est Real HRT], [Sx2 Est Real HRT], [Sx3 Est Real HRT], [Sx4 Est Real HRT], [Sx5 Est Real HRT] and [Sx6 Est Real HRT]. Null values indicate that the compartment does not exist, and are therefore ignored.
Tot . Stages	Total number of stages, equivalent to [No of Sltr Stages] + [No of AB Stages]
% RAS Flow	[RAS Flow (MGD)] / [Flow (MGD)]

Field	Description
Tot Reac Vol (MG)	Total reactor volume in service, equivalent to [Tot Sltr Vol in Svc (MG)] + [Tot AB Vol (MG)] +[Extra Vol (MG)]. Usually [Extra Vol (MG)] will be a null value.
Sltr Vol Frac	The fraction of the total reactor volume occupied by the selector compartments, equivalent to [Tot Sltr Vol in Svc (MG)] / [Tot Reac Vol (MG)]
Sx1 F/M (lb BOD_5/lb MLSS-d)	F/M ratio calculated as [Flow (MGD)]*[AS Inf BOD_5 (mg/L) calc]/([Sx1 Vol in Svc (MG)]*[Avg MLSS (mg/L) calc]). For the Davenport WPCP (#056), the contact zone MLSS was used instead of the average MLSS.
Sx1 HRT (h)	The nominal HRT (h) within Sx1, calculated as [Sx1 Vol in Svc (MG)]/[Flow (MGD)]*24. This nominal HRT value does not take into account RAS recycle and mixed liquor recycle flows (if present), and therefore are an over-estimate of the actual HRT.
Sx1 HRT incl RAS (h)	The HRT (h) within Sx1, calculated as [Sx1 Vol in Svc (MG)]/[Flow+RAS (MGD)]*24. This HRT value accounts for RAS recycle, but not for mixed liquor recycle flows (if present). For plants that do not practice mixed liquor recycle, this can be considered the actual HRT.
Tot Sltr F/M (lb BOD_5/lb MLSS-d)	Same as [Sx1 F/M (lb BOD_5/lb MLSS-d)], except calculated using the entire selector volume, [Tot Sltr Vol in Svc (MG)].
Tot Sltr HRT (h)	The nominal HRT (h) within all selector compartments calculated as [Tot Vol in Svc (MG)]/[Flow (MGD)]*24. This nominal HRT value does not take into account RAS recycle and mixed liquor recycle flows (if present), and therefore are an over-estimate of the actual HRT.
Tot Sltr HRT incl RAS (h)	The HRT (h) within all selector compartments, calculated as [Tot Vol in Svc (MG)]/[Flow+RAS (MGD)]*24. This HRT value accounts for RAS recycle, but not for mixed liquor recycle flows (if present). For plants that do not practice mixed liquor recycle, this can be considered the actual HRT.
Sltr MCRT (d)	Selector MCRT calculated on a daily basis (i.e., using wasting data and solids inventory for each day). Clarifier solids and effluent suspended solids are not included in the calculation. The calculation formula is [Tot Sltr Vol in Svc (MG)]*[Avg MLSS (mg/L) calc]*8.34/([Wastage (lb SS/d)]. Values above 10 are reported as 10. On days where no wasting took place, this value is not reported.
7d Avg Sltr MCRT (d)	An MCRT value based on average solids wasting over the last 7 days. The calculation formula is [Tot Sltr Vol in Svc (MG)]*[Avg MLSS (mg/L) calc]*8.34/([7d_Wastage(lb SS/d)], where [7d_Wastage(lb SS/d)] is an average of [Wastage (lb SS/d)] over the last 7 days. The field [7d_Wastage(lb SS/d)] is for internal calculation only and is not provided. Note that [7d Avg Sltr MCRT (d)] is not an average of the [Sltr MCRT (d)]

Field	Description
	value over the last 7 days, which would give excessive weight to days where solid wasting was minimal (and therefore [Sltr MCRT (d)] was very high).
7d Sltr MCRT > 1	1 if [7d Avg Sltr MCRT (d)] is less than 1. 0 if greater than or equal to 1.
7d Sltr MCRT > 2	1 if [7d Avg Sltr MCRT (d)] is less than 2. 0 if greater than or equal to 2.
7d Sltr MCRT > 3	1 if [7d Avg Sltr MCRT (d)] is less than 3. 0 if greater than or equal to 3.
Reactor MCRT (d)	Reactor MCRT calculated on a daily basis (i.e., using wasting data and solids inventory for each day). Clarifier solids and effluent suspended solids are not included in the calculation. The calculation formula is [Tot Reac Vol (MG)]*[Avg MLSS (mg/L)]*8.34/([Wastage (lb SS/d)]. Values above 50 are reported as 50. On days where no wasting took place, this value is not reported.
7d Avg Reactor MCRT (d)	An MCRT value based on average solids wasting over the last 7 days. The calculation formula is [Tot Reac Vol in Svc (MG)]*[Avg MLSS (mg/L) calc]*8.34/([7d_Wastage(lb SS/d)], where [7d_Wastage(lb SS/d)] is an average of [Wastage (lb SS/d)] over the last 7 days. The field [7d_Wastage(lb SS/d)] is for internal calculation only and is not provided. Note that [7d Avg Reactor MCRT (d)] is **not** an average of the [Reactor MCRT (d)] value over the last 7 days, which would give excessive weight to days where solid wasting was minimal (and therefore [Reactor MCRT (d)] was very high).
Sx1 Aer	1 if Sx1 is aerobic, 0 if it is not. For most plants, Sx1 was classified as aerobic, anoxic, or anaerobic and this classification was applied throughout the study period. Two plants, Renton (#010) and UOSA (#126) have data sets where the classification of the selector changed during the study period.
Sx1 Anx	1 if Sx1 is anoxic, 0 if it is not. For most plants, Sx1 was classified as aerobic, anoxic, or anaerobic and this classification was applied throughout the study period. Two plants, Renton (#010) and UOSA (#126) have data sets where the classification of the selector changed during the study period.
Sx1 Anb	1 if Sx1 is anaerobic, 0 if it is not. For most plants, Sx1 was classified as aerobic, anoxic, or anaerobic and this classification was applied throughout the study period. Two plants, Renton (#010) and UOSA (#126) have data sets where the classification of the selector changed during the study period.

Field	Description
pH calc	pH value measured in the aeration basin (9 data sets), or in the secondary or plant effluent (30 data sets). "Calc" indicates that the data includes estimated values as substitutes for missing values, as described separately in Section 4.3.2.6.
Sltr DO (mg/L)	DO concentration measured in one of the selector compartments. This field is not well populated, as DO is often not monitored in anaerobic or anoxic zones.
AB DO (mg/L) calc	DO concentration measured in the aeration basin. For three plants, DO concentration data from the aeration basin was unavailable, and secondary effluent DO concentration was substituted. "Calc" indicates that the data includes estimated values as substitutes for missing values, as described separately in Section 4.3.2.6.
RAS Cl2	1 indicates that RAS chlorination was used at this plant on this day (if usage dates are known) or RAS chlorination was used at this plant (or likely used) during the study period but the exact dates of usage are unknown. 0 indicates that RAS chlorination was not being used at this plant this day (if usage dates are known), but was used during study period or RAS chlorination was not used at this plant at any time during the study period.
Filament_Type_MCRT	Indicates whether the dominant filament(s) at the plant tend to be short (value = 0) or long (value = 1) MCRT filaments, or are likely to be based on calculated MCRT and whether or not nitrification is practiced.
Contact MLSS (mg/L) calc	MLSS concentration in the contact zone. Used only for Davenport WPCP (#056), a contact-stabilization plant. "Calc" indicates that the data includes estimated values as substitutes for missing values, as described separately in Section 4.3.2.6.

APPENDIX D

FURTHER DISCUSSION OF THE REGRESSION ANALYSES

"Statistics is a way to get information from data."

Keller and Warrack, 2000

There may be some concern regarding the low R^2 values reported in this study report. Many of us in the wastewater treatment field are used to seeing R^2 values from controlled laboratory experiments, which are often 90% and higher. We sometimes hear from our colleagues and associates that relationships below some arbitrary R^2 value are invalid. Maybe this cut off value is 70% or 60% or even 50%. Although no one is able to find this cut off value in any reputable reference, we continue to believe one exists. The existence of a cut off value, however, is a myth and demonstrates a misunderstanding of the R^2 statistic, and possibly regression analysis in general.

The coefficient of determination, R^2, measures that proportion of the variation in the dependent variable (log DSVI in our study) that is explained by the variation in the independent variable(s) (selector design/operating parameters), and thus R^2 is used to compare the strength of different regression models (Keller and Warrack, 2000). For example, $R^2 = 4.8$ for the number of aeration basin stages regressed against log DSVI in the long-MCRT group. This means that the number of aeration basin stages in long-MCRT plants accounts for only 4.8% of the variation in DSVI or sludge settleability. If we only consider the number of aeration basin stages in a long-MCRT plant, we would miss the remaining 95.2% of the variability in DSVI. Using the R^2 value to compare the influence that average MLSS had on DSVI to the influence that the number of aeration basin stages had on DSVI, we see that the average MLSS with an $R^2 = 23.4$ had substantially more influence on DSVI than the number of aeration basin stages had on DSVI.

Adding more design/operating parameters (independent variables) to the regression analysis will reduce the portion of DSVI variation that is not explained by the regression analysis, and therefore the R^2 value will increase. For the long-MCRT group, 7 design/operating parameters were included in a multiple regression analysis (Table D-1). The R^2 for the multiple regression analysis was 42.3%, or almost double the highest single regression variable R^2 (average MLSS $R^2 = 23.4\%$) from the same dataset. Even the multiple regression analysis, however, accounts for less than half of the variation in DSVI.

Reisinger, 1997, investigated how various research designs could impact R^2 values and found that data type had a significant effect on R^2. Average R^2 values for time series, cross-sectional, and pooled data (combination of both time series and cross-sectional data) were 60%, 31%, and 52%, respectively. He found that the main difference between time series data and cross-sectional data was that time series data was aggregated where cross-sectional data was not. Reisinger concluded that some unexplained variation was averaged out of the aggregated data, where it was not in the unaggregated cross-sectional data. Since our datasets for all three plant groups were unaggregated daily operating data, we should expect lower R^2 values per Reisinger. Based on Reisinger's work, the multiple regression R^2 values in our study were higher than the average R^2 values in the unaggregated cross-section data studies.

Table D-1. Long-MCRT Plant Group Multiple Regression Analysis.

Predictor	T-statistic	P-value	VIF
Constant	305.60	0.000	
Primary Treatment? (Yes/No)	-9.59	0.000	1.2
Average MLSS (mg/L)	-54.71	0.000	1.2
Effluent Temperature (°C)	-30.27	0.000	1.5
No. of Selector Stages	22.73	0.000	1.5
No. of Aeration Basin Stages	-11.37	0.000	1.6
Selector Volume/Total Basin Volume	12.41	0.000	1.2
Aeration Basin DO (mg/L)	-7.80	0.000	1.3

S = 0.1037 R-Sq = 42.3% R-Sq(adj) = 42.3%

Analysis of Variance

Source	DF	SS	MS	F	P
Regression	7	58.9597	8.4228	783.30	0.000

Our complete dataset is composed of full-scale data collected at 44 different wastewater treatment facilities, with 44 different laboratory staffs, 44 different operations and maintenance staffs, with variation of staff expertise, methods used to collect and analyze samples, methods used to take measurements, variation in meter calibration and other measuring instruments' calibration, diurnal variation patterns in temperature, pH, sewage characteristics, etc., which were not captured by the data, could introduce significant variation between plants that was not included in the regression analyses, and as a result probably lowered the R^2 value substantially. Other parameters that were only measured by a few plants (so were not included in the regression analysis) such as: nitrate and nitrite concentrations in anoxic selectors, and orthophosphate concentration in anaerobic selectors, may also account for the variation in DSVI not accounted for in the regression analyses.

The design/operating parameters may not have been analyzed in the proper form. The single regression analyses in this study and the multiple regression analysis in Table D-1, assumed that the relationship between DSVI and the design/operating parameters was linear. In Table 4-14, the linear R^2 value is only 1.1% for the effective nominal ICZ HRT parameter, but the cubic polynomial regression R^2 is 7.4%, for the same parameter. Even the cubic polynomial R^2 is low, but is still almost seven times as high as the linear R^2.

If the R^2 values are so low, how do we know a relationship between DSVI and the design/operating parameter exists at all? If the t-statistic is greater than 2.0 or less than -2.0 (i.e., the absolute value of the t-statistic is greater than 2.0), then there is a significant relationship between the design/operating parameter and DSVI (DeLurgio, 1998). Further, if the p-value is less than 1%, there is "overwhelming evidence" that the regression relationship between the design/operating parameter and DSVI is valid and highly significant (Keller and Warrack, 2000); regardless of the R^2 value. All the parameters that we present as "significant" had absolute value t-statistics much greater than 2.0 and p-values = 0.000 (or ≤ 0.0%).

T-statistics were higher in the long-MCRT group than in the other groups, and higher in the short-MCRT group with anaerobic and anoxic selectors compared to the short-MCRT group with aerobic selectors, for a given R^2 value. The long-MCRT group also had many more data points for each parameter (approx. 9,000) than the other two groups, and the short-MCRT group with anaerobic and anoxic selectors had almost five times as many data points per parameter than the short-MCRT group with aerobic selectors. This demonstrates how larger sample sizes increase t-statistic values.

A Type I error is when we reject a true hypothesis, and a Type II error is when we do not reject a false hypothesis. By increasing the sample size, we reduce the probability of a Type II error and strengthen the validity of the regression analysis (Keller and Warrack, 2000). With larger datasets used, the regression analysis becomes more precise. The very large sample size used in the long-MCRT group provides strong support for the outcomes of the regression analysis. Although 1,000 data points still provides strong support for the short-MCRT with aerobic selectors group's regression results, the results might not be as strongly supported as the regression results from the long-MCRT plant group.

Food-to-Microorganism Ratio (F/M) and Kinetic Reaction Rates

In contrast to the literature, the regression analyses in this study do not strongly correlate initial contact zone (ICZ) F/M with enhanced selector kinetics. Although the short-MCRT with aerated selectors group regression analysis showed that DSVI decreased with increased ICZ F/M, the $R^2 = 9.4\%$ was much lower than that for the ICZ HRT with an $R^2 = 33.7\%$. Kinetic reaction rates are generally a function of reactant concentration and temperature (Barrow, 1973). The higher the BOD concentration and temperature, the faster the BOD consumption rate should be. It also follows that the higher the active microorganism concentration (measured with MLSS), the higher the BOD consumption rate should be. Since the ICZ F/M includes both of these parameters (BOD and MLSS), could we conclude that ICZ F/M should provide the best predictor of BOD uptake rates in a selector?

Since F/M specifically measures the loading of BOD on the MLSS, and not the actual concentration of either BOD or MLSS, the F/M may not be a good predictor of BOD uptake rates. For a given F/M, the concentration of BOD and MLSS can be very high or very low. For example, using the following formula:

$$F/M = \frac{\text{BOD concentration (mg/L)} \times \text{sewage flow rate (L/d)}}{\text{MLSS concentration (mg/L)} \times \text{basin volume (L)}}$$

If the ICZ F/M = 6.0 kg BOD/kg MLSS-d, and for simplicity the sewage flow rate is 10^6 L/d, and the basin volume is 10^5 L; then the BOD concentration/MLSS concentration (BOD/MLSS) = 0.6. This means that the BOD concentration would be 3,000 mg/L, for a MLSS = 5,000 mg/L; or a BOD = 90 mg/l for a MLSS = 150 mg/L. Obviously the kinetic rates will be much greater in the first case than that in the second case. The ICZ F/M, however, is the same in both cases. Therefore, the ICZ F/M is not a good predictor of kinetic reaction rates, including BOD uptake rates in a selector. In general, however, as the ICZ HRT decreases, the higher the BOD concentration will be in the ICZ. Therefore, the ICZ HRT may be a better predictor of kinetic rates than the ICZ F/M. This is consistent with the regression analysis results for the short-MCRT plants with aerobic selectors (ICZ HRT $R^2 = 33.7\%$--see Table 4-13), where the activated sludge influent BOD concentration had the most influence on DSVI ($R^2 = 36.7\%$).

Since the F/M is a measure of BOD load on MLSS, the F/M should predict whether a mixed liquor will be overwhelmed or not by the amount of BOD entering the selector. Therefore, the F/M should be a good predictor of when BOD leakage to the main aeration basin might occur. The lower the selector F/M, the lower the risk of BOD leakage would be. This is consistent with the regression results for the short-MCRT plants with anaerobic or anoxic selectors (see Table 4-13), where the lowest selector F/Ms are associated with the lowest DSVIs.

The selector literature provides good support for the results of the regression analyses in this study. The regression statistics strongly support the conclusions derived from this study. Using simple plots of average plant values and 95 percentile DSVIs did not provide much useful information about selector design/operating parameters. The regression analysis, however, provided a wealth of useful information. This fully supports Keller and Warrack's very simple but poignant statement: *"Statistics is a way to get information from data."*

References:

Barrow, G.M. (1973) Physical Chemistry 3rd edition. McGraw Hill, Inc.

DeLurgio, S.A. (1998) Forecasting Principles and Applications. McGraw Hill, Inc.

Eisinger, H. (1997) *The impact of research designs on R^2 in linear regression models: an exploratory meta-analysis.* Journal of Empirical Generalizations in Marketing Science, Volume 2.

Keller, G., Warrack, B (2000) Statistics for Management and Economics. Duxbury, Thomas Learning.

APPENDIX E

PERCENTILE DISTRIBUTION ANALYSIS OF REGRESSION ANALYSIS DATASETS

Percentile distributions were calculated for key selector design and operating parameters to further compare each dataset used in the regression analysis presented in Chapter 4.0 (i.e., long-MCRT plants, short-MCRT plants with anaerobic or anoxic selectors, and short-MCRT plants with aerated selectors). Figure E-1 shows that the DSVIs for the long-MCRT plants were significantly lower than the DSVIs for the short-MCRT plants. In fact, the long-MCRT plants had DSVIs ≤93 mL/g 50% of the time, compared to the short-MCRT plants with anaerobic or anoxic selectors, which had DSVIs ≤133 mL/g 50% of the time, and the short-MCRT plants with aerobic selectors, which had DSVIs ≤110 mL/g 50% of the time.

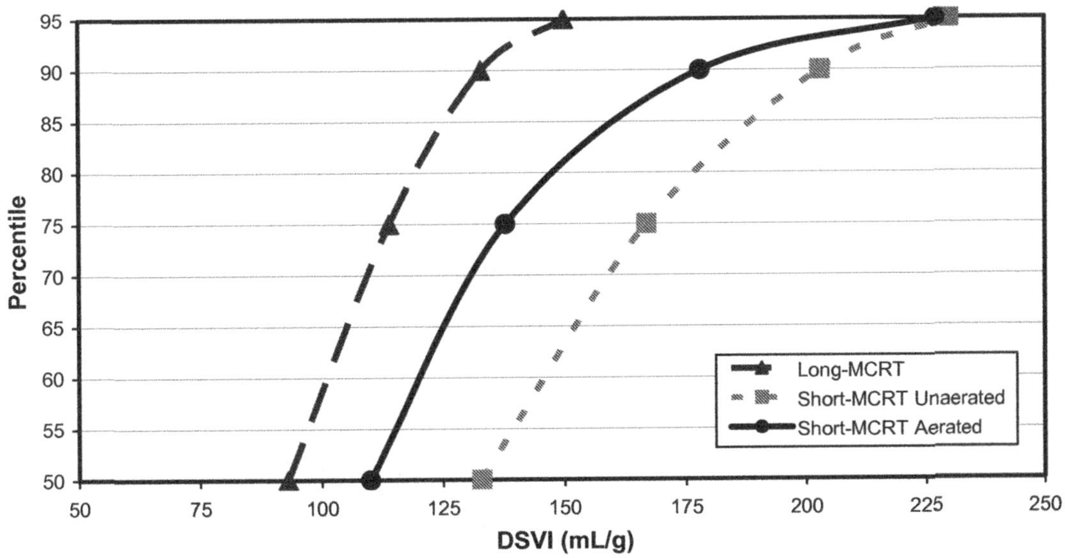

Figure E-1. Percentile Distribution of DSVI for the Three Data Sets.

This suggests that selectors in long-MCRT plants were more successful in controlling filamentous bulking than selectors in short-MCRT plants, but Wanner (1994) offers another explanation. Wanner distinguishes between filamentous organisms (genus or type) that can cause severe bulking (SVIs >200–300 mL/g) and those that typically cause less severe bulking (SVIs do not increase above 200–300 mL/g) in activated sludge. Filamentous organisms that tend to cause higher SVIs were mostly those that typically dominate short-MCRT activated sludges (e.g., Type 021N, *Thiothrix*, and *Sphaerotilus natans*), while the filamentous organisms that tend to cause lower relative SVIs were mostly those that dominate long-MCRT activated sludges (e.g., *Microthrix parvicella*, *Haliscomenobacter hydrossis*, Type 0092, and *Nostocoida limicola*). Therefore, the lower DSVI found in long-MCRT plants may be because of the type of filamentous organisms that grow in these activated sludges rather than because the short-MCRT selectors were effective.

Figure E-2 shows that compared to DSVI, SVI was not as different between datasets, although the long-MCRT plants still had the lowest SVIs. Wanner (1994) showed that municipal wastewater treatment plants (presumably without selectors) typically had SVIs ≤103 mL/g 50% of the time and SVIs ≤148 mL/g 84% of the time. This compares reasonably well with the long-MCRT plants that had SVIs ≤106 mL/g 50% of the time, and ≤162 mL/g 84% of the time. SVIs were ≥150 mL/g 21% of the time for long-MCRT plants, 40% of the time for short-MCRT plants with anaerobic or anoxic selectors, and 25% of the time for short-MCRT plants with aerobic selectors.

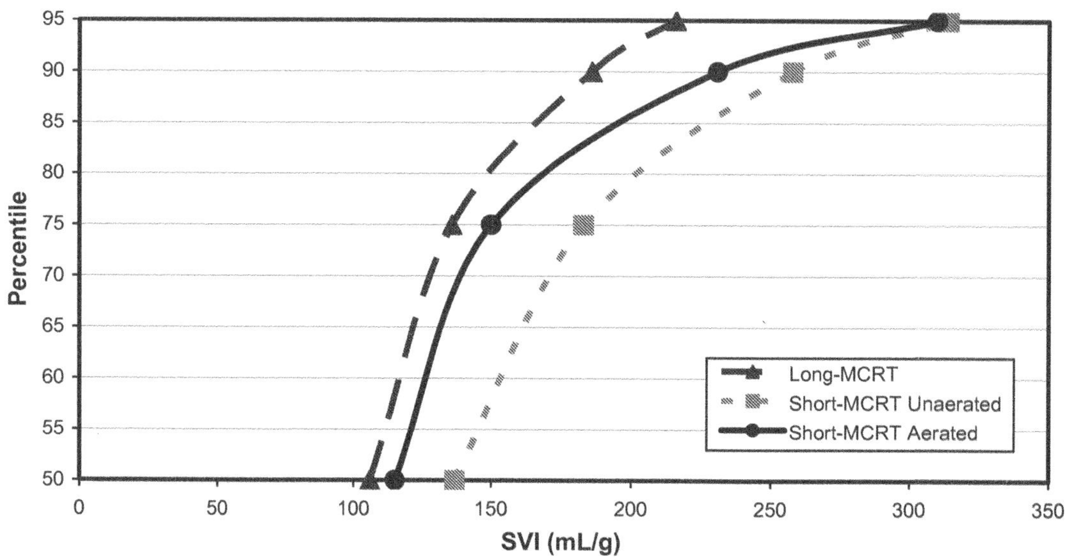

Figure E-2. Percentile Distribution of SVI for the Three Datasets.

Figure E-3 shows that the ICZ F/M is lowest for the long-MCRT plants with ICZ F/Ms ≤1.9 kg BOD_5/(kg MLSS·d) 50% of the time, compared to short-MCRT plants with anaerobic or anoxic selectors, which had ICZ F/Ms ≤3.6 kg BOD_5/(kg MLSS·d) 50% of the time, and short-MCRT plants with aerobic selectors, which had ICZ F/Ms ≤8.9 kg BOD_5/(kg MLSS·d) 50% of the time. Since higher ICZ F/Ms [at least >3 kg BOD_5/(kg MLSS·d), per Jenkins et. al., 2004 and Wanner, 1994] are usually recommended to produce a sufficient BOD_5 concentration gradient in anoxic or anaerobic selectors to control bulking, this suggests that ICZ F/M was not the reason that the long-MCRT plants had lower SVIs than the short-MCRT plants. This supports similar findings discussed in Chapter 4.0.

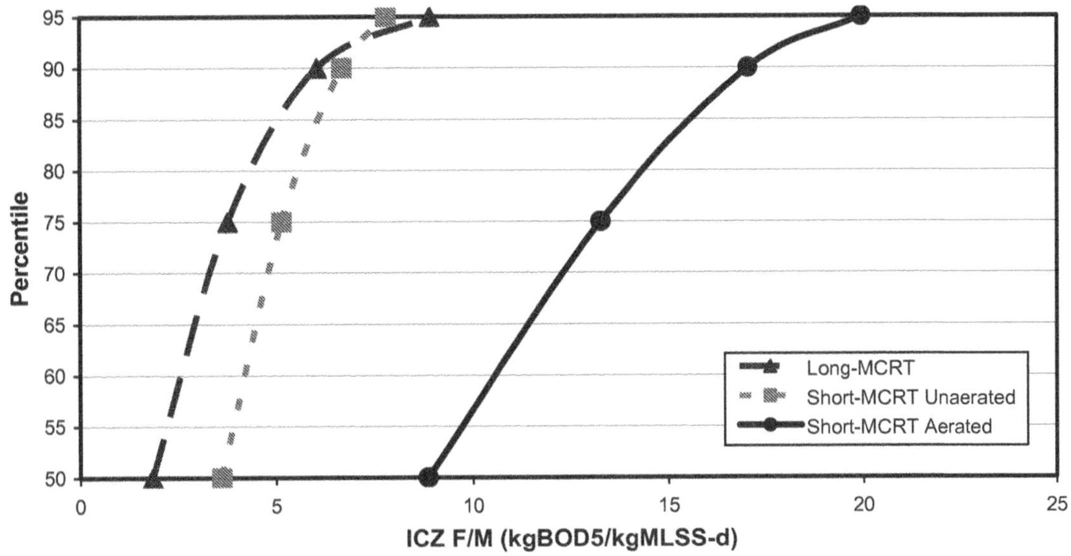

Figure E-3. Percentile Distribution of ICZ F/M for the Three Datasets.

Figure E-4 shows that the activated sludge influent BOD_5 concentration was significantly higher in the long-MCRT plants compared to the short-MCRT plants. Figure E-5 shows that the selector influent BOD_5/TSS ratio is similar (just above 1.0) for the long-MCRT and short-MCRT plants at least 50% of the time, but the long-MCRT plants have a wider range of BOD_5/TSS values than the short-MCRT plants with anaerobic or anoxic selectors. Both of these plant groups have a much wider range of BOD_5/TSS values than the short-MCRT plants with aerobic selectors, which have a much smaller number of plants in the dataset. Since the BOD_5/TSS ratio is an indication of soluble BOD_5 concentration (or raCOD), the long-MCRT plants do not appear to impacted by filamentous bulking when soluble BOD_5 concentrations are high in their feed, since DSVIs in these plants are low even when the BOD_5/TSS ratio and influent BOD_5 concentration may be high.

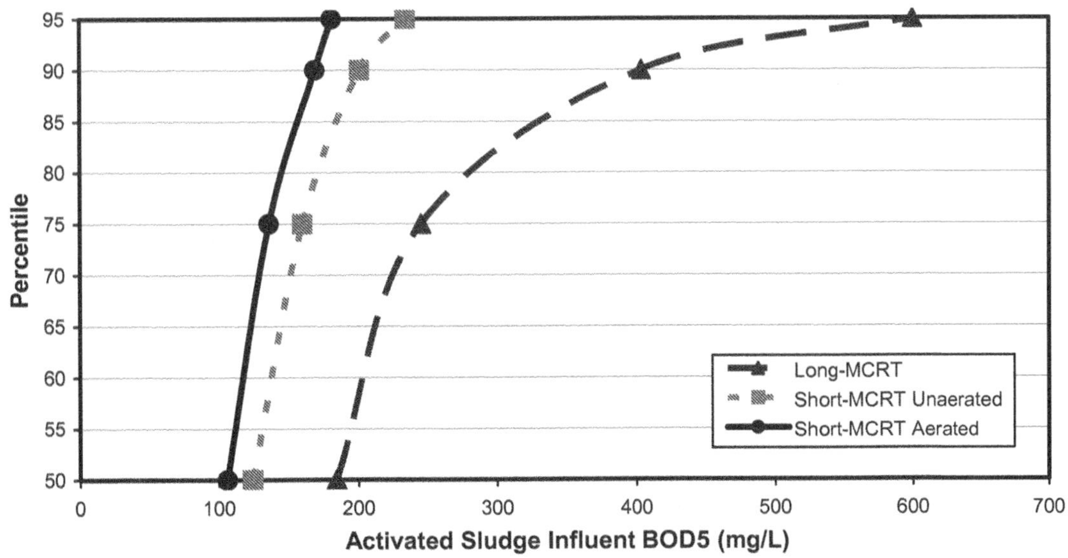

Figure E-4. Percentile Distribution of Activated Sludge Influent BOD_5 Concentration for the Three Datasets.

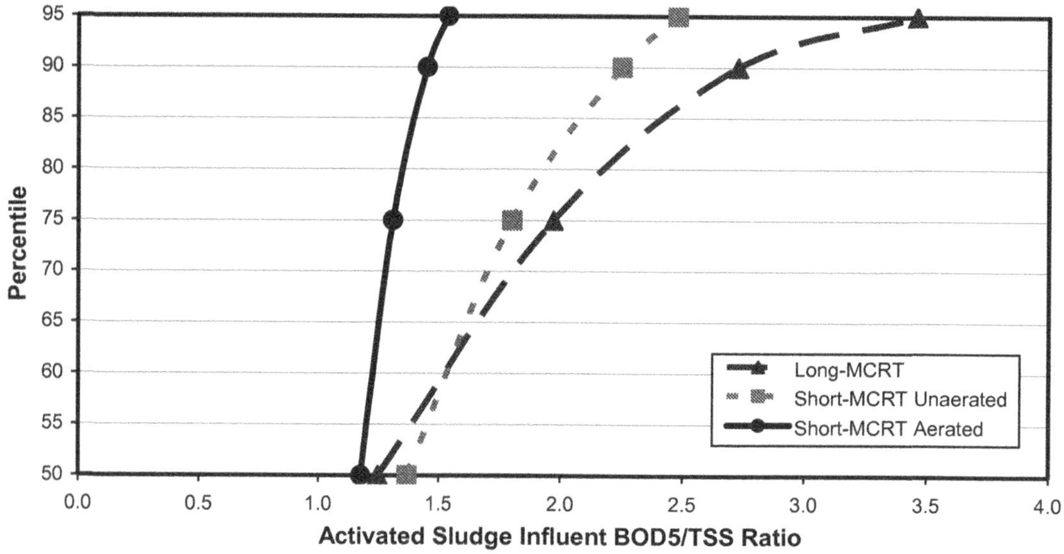

Figure E-5. Percentile Distribution of Activated Sludge Influent BOD_5/TSS Ratio for the Three Datasets.

Although the influent BOD_5 is the highest for long-MCRT plants, Figure E-6 shows that the long-MCRT plants have the lowest selector F/M. Figure E-7 shows that the long-MCRT plants also tend to have longer HRTs, which explains why the F/M is lowest for long-MCRT plants. Figure E-7 also shows that long-MCRT plant HRTs have a wider range of values than the short-MCRT plants.

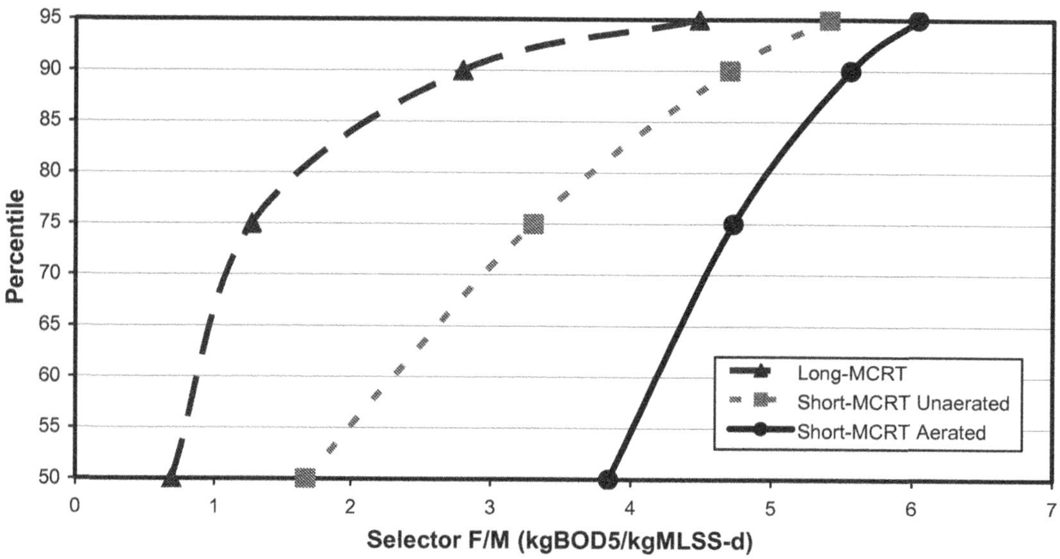

Figure E-6. Percentile Distribution of Selector F/M for the Three Datasets.

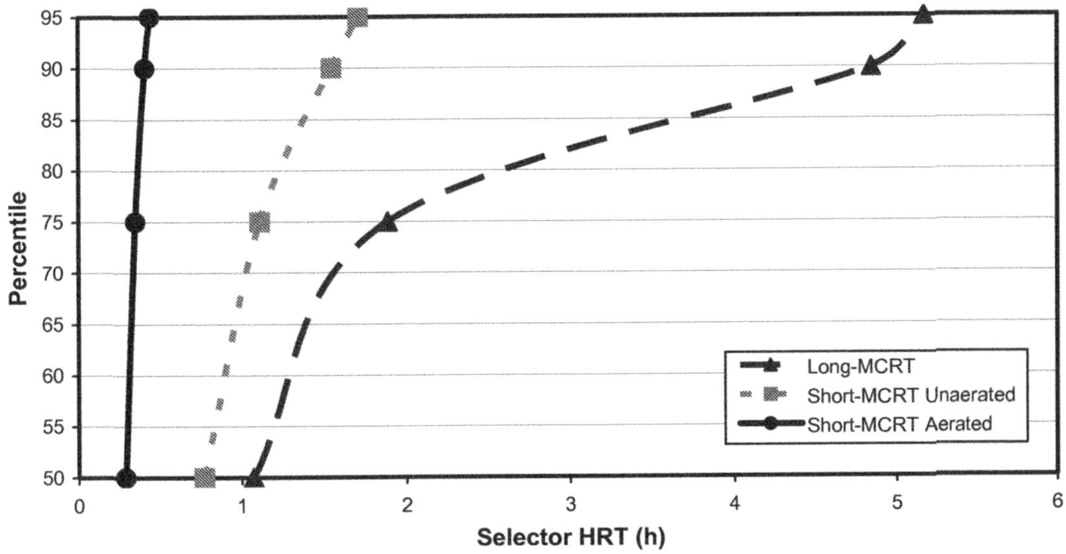

Figure E-7. Percentile Distribution of Selector HRTs for the Three Datasets.

Figure E-8 shows that long-MCRT plant aeration basin DO was higher than that in short-MCRT plants with anaerobic or anoxic selectors. The short-MCRT plants with aerobic selectors were mainly high-purity oxygen activated sludge plants with aeration basin DO concentrations that are not comparable to air activated sludge plants; therefore, these plants were not included in Figure E-8.

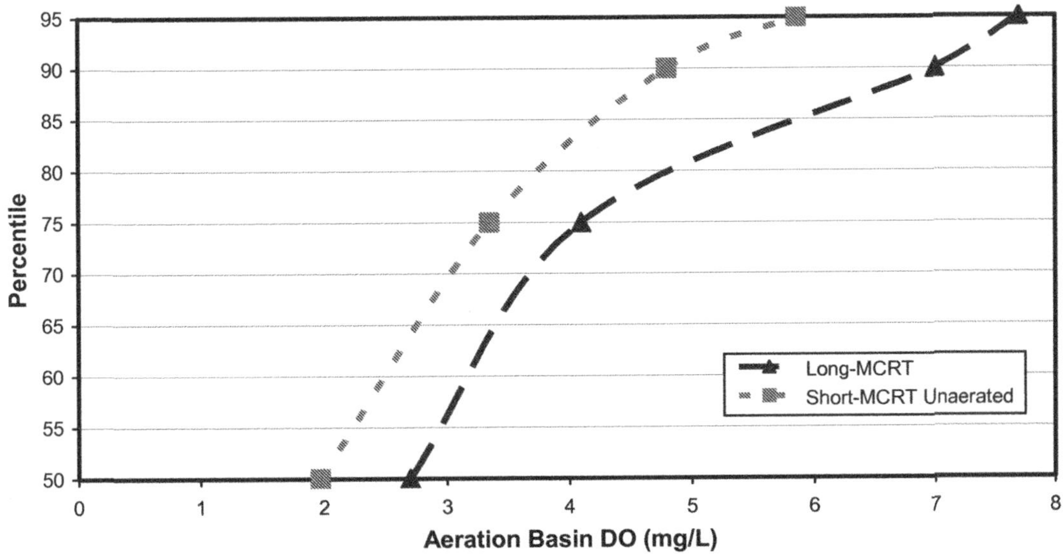

Figure E-8. Percentile Distribution of Aeration Basin Dissolved Oxygen Concentration for the Three Datasets.

APPENDIX F

INSTRUCTIONS FOR SELECTOR DIAGNOSTIC TOOL

Introduction

The Selector Diagnostic Tool, included on the CD attached to the inside back cover of this report, is intended to assist operators who have selector systems and are experiencing problems with sludge settleability or engineers who are designing selector systems. Settleability problems are characterized by high DSVIs or SVIs.

The application asks the user to provide information about your system, and draws upon the study results and literature review in order to provide the user with suggestions that may improve sludge settleability.

In order to use the application, you will need to provide the following information to categorize your plant:

- **Dominant Filament Type(s)** – Filament types are typically identified by microscopic analysis during bulking episodes. As different filaments have different growth requirements, your system will be categorized as long-MCRT or short-MCRT based on the dominant filament type(s) that you identify. If you don't know the dominant filament type(s), you will be asked to enter the total basin MCRT instead.
- **Selector Type** – Selectors are categorized as aerobic, anoxic or anaerobic. If your selector has multiple compartments of differing types, enter the type for the first compartment, or initial contact zone.

In addition, you will be asked to provide additional plant information, such as the number of selector stages, average mixed liquor suspended solids (MLSS), food-to-microorganism (F/M) ratios, hydraulic retention times (HRTs), temperature and aeration basin dissolved oxygen (DO) concentration. These parameters have been found to be significantly correlated with sludge settleability (DSVI). Refer to Appendix C for details about how each parameter is calculated or obtained.

Start the Application

The application is contained in an Access database file, named **selector_diagnostic.mdb**. Upon opening the file, you may receive a general warning message, "This file may not be safe if it contains code that was intended to harm your computer. Do you want to open this file or cancel the operation?" Click open to continue using the application.

Select Dominant Filament(s)

Select the dominant filament(s) in your mixed liquor during bulking episodes, based on microscopic analysis. Select multiple filaments by holding down the CTRL key while making your selections. If you select dominant filaments that are typical of both long-MCRT and short-MCRT plants, you will be asked to enter additional plant information for both types of plants, and you will receive a set of results for each type.

If you don't know the type of filaments in your mixed liquor, scroll down and select "Don't know." You will be asked instead to enter the total basin MCRT for your system (excluding clarifier solids).

Example: Select "Type 1701" and click the CONTINUE button (Figure F-1).

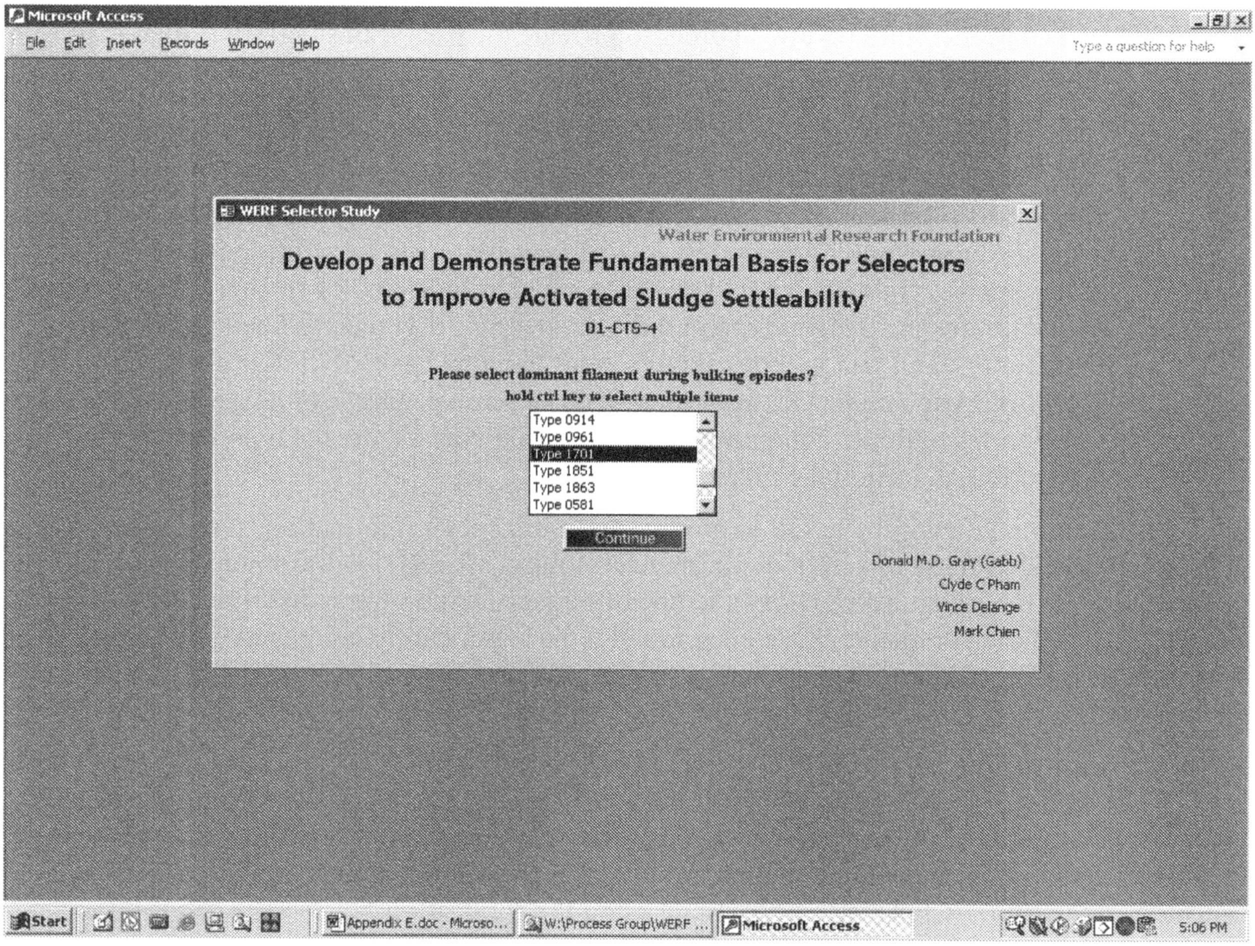

Figure F-1. Select Dominant Filament Type(s).

Identify Selector Type

Choose a selector type (aerobic, anoxic or anaerobic) that best represents the conditions in the first selector compartment, or initial contact zone (ICZ).

Example: Select "Anaerobic" and click the CONTINUE button (Figure F-2).

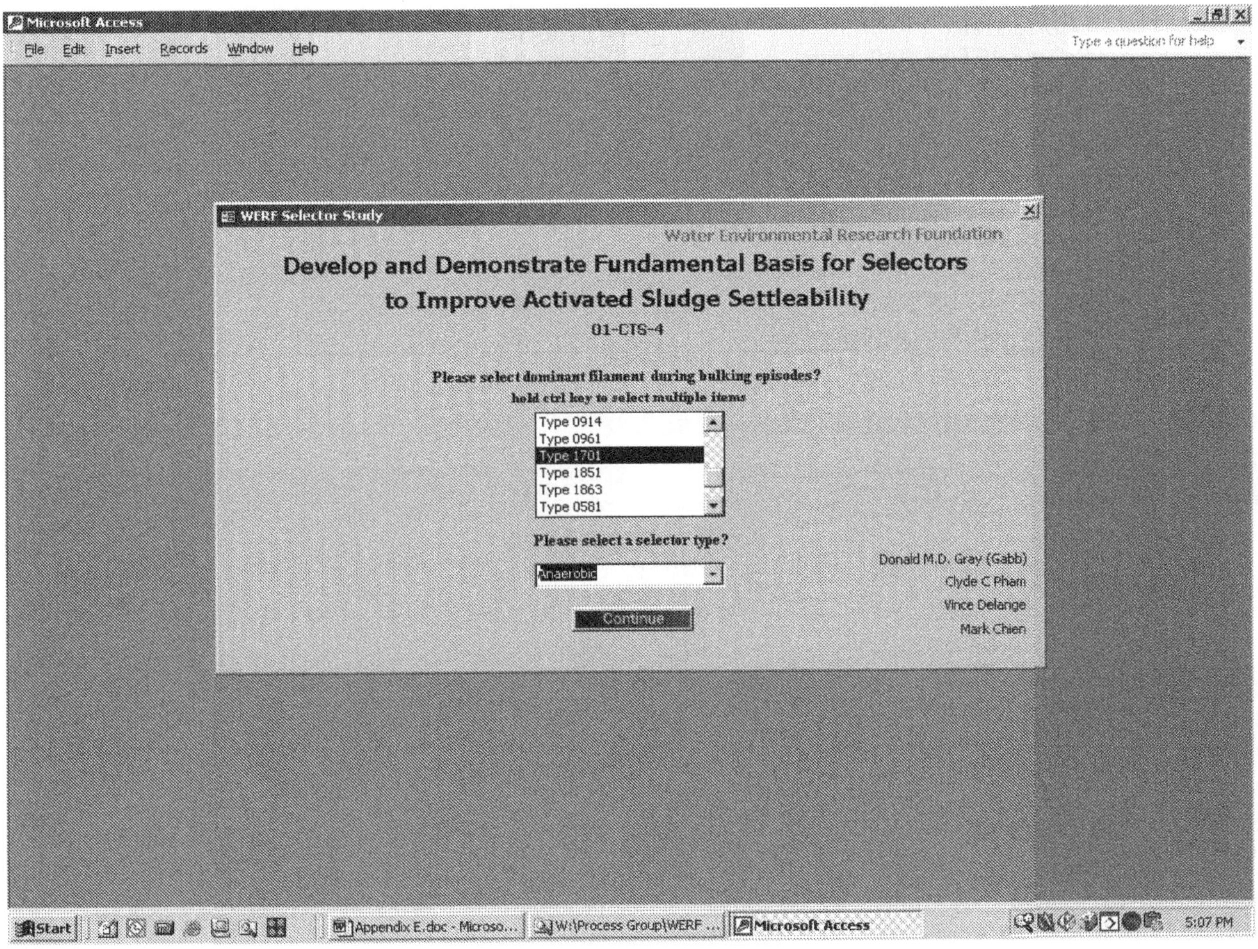

Figure F-2. Identify Selector Type.

Enter Additional Plant Information

Based on the filament type(s) and selector type, you will be prompted to provide additional plant operating conditions that may affect your sludge settleability. For each field, select the range that best applies to your system. Leave the field blank if the information is unavailable.

Refer to Appendix C for details about how each parameter is calculated or obtained.

Example: Enter the plant information in the Range Selected column below and click the SUBMIT DATA button (Figure F-3):

PARAMETER	EXAMPLE PLANT VALUE	RANGE SELECTED
Average MLSS (mg/L)	1400	< 1500
Total Selector F/M (kg BOD5/kg MLSS-days)	1.9	> 1
No. of Selector Stages	1	1
BOD/TSS ratio	1.2	1.0 – 2.0
Selector HRT w/o RAS (hours)	1.4	1.2 – 2.5
Selector HRT w/ RAS (hours)	0.7	< 0.75
ICZ HRT w/o RAS (hours)	1.4	< 2.4
Reactor MCRT (days)	3.5	1.5 – 4.5
Selector MCRT (days)	0.5	<1.0
Aeration Basin DO (mg/L)	3.5	2.5 – 4.0
ICZ F/M (kg BOD5/kg TSS-days)	1.9	1.0 – 3.0
Selector Vol/ Total Basin Vol Ratio (%)	14%	< 22.5%
ICZ HRT w/ RAS (hours)	0.7	< 1.4
Effluent Temperature (deg Celsius)	22	20 – 25

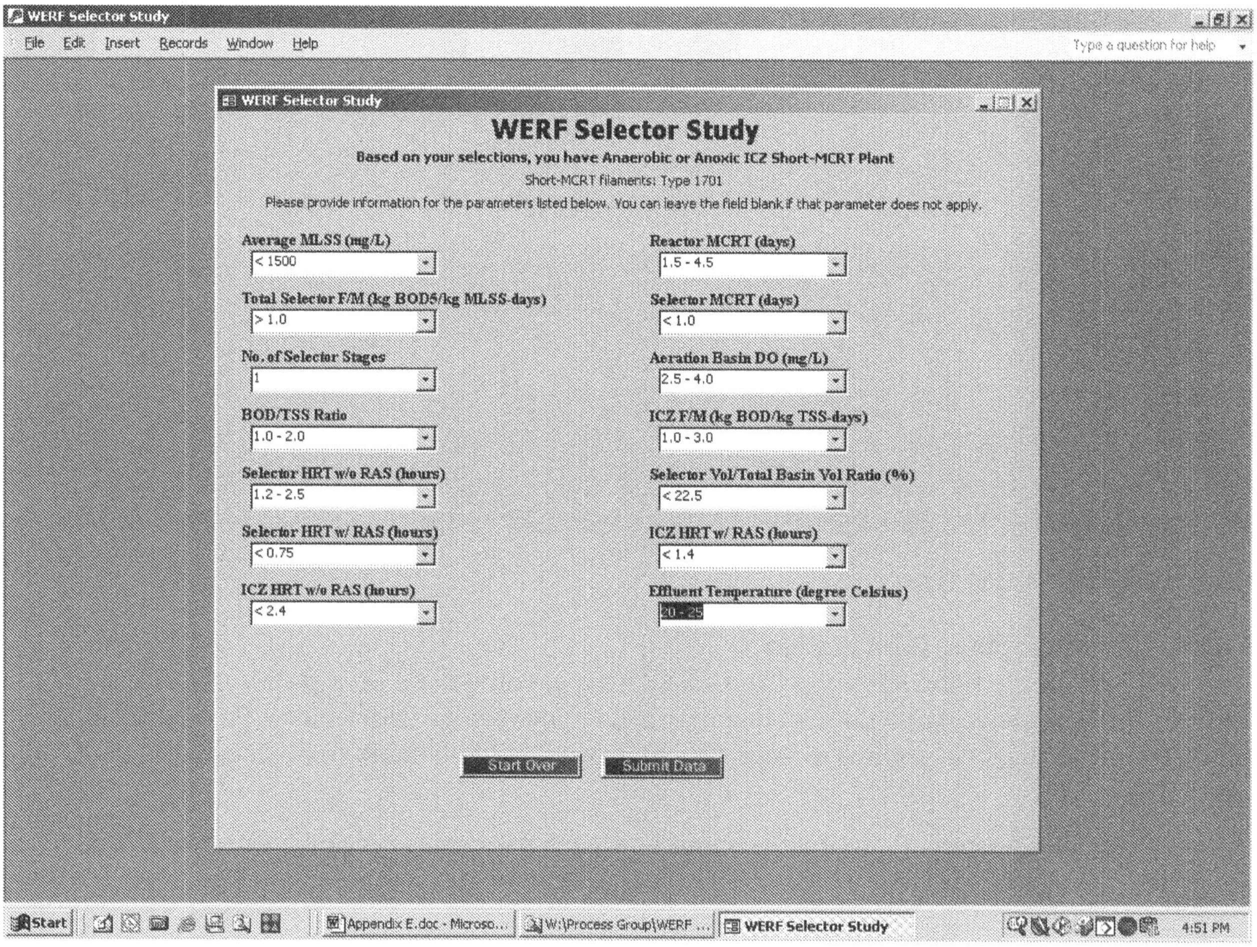

Figure F-3. Enter Additional Plant Information.

Results

The main results window (Figure F-4) contains a table that shows:

- A list of plant operating parameters that may affect settleability,
- Values selected for your plant,
- Recommended values based on the regression analysis from this study,
- Recommended values from the literature, and
- References for the recommended literature values.

Parameters that are out of the recommended ranges are highlighted in red.

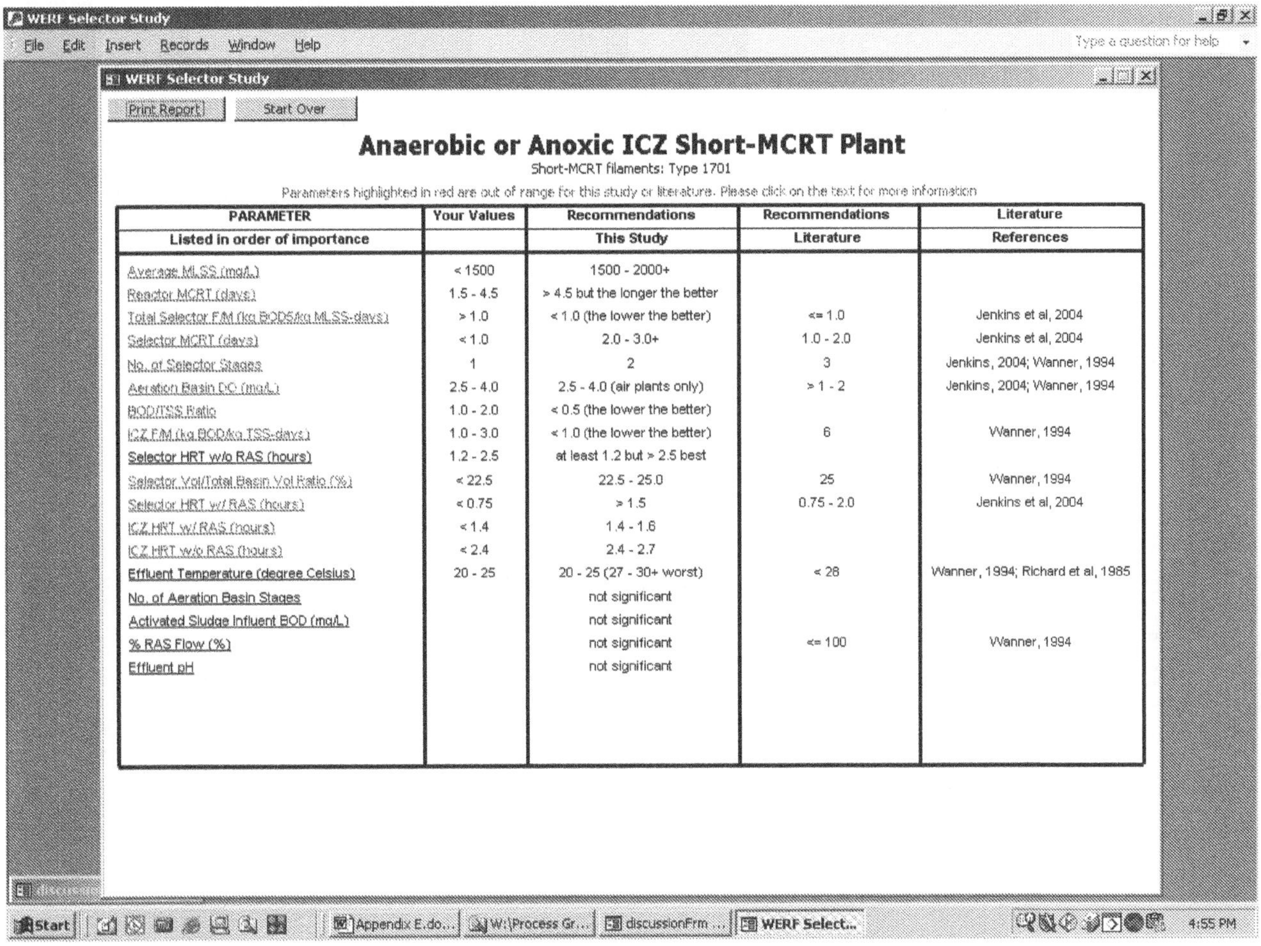

WERF Selector Study

Anaerobic or Anoxic ICZ Short-MCRT Plant

Short-MCRT filaments: Type 1701

Parameters highlighted in red are out of range for this study or literature. Please click on the text for more information

PARAMETER	Your Values	Recommendations	Recommendations	Literature
Listed in order of importance		This Study	Literature	References
Average MLSS (mg/L)	< 1500	1500 - 2000+		
Reactor MCRT (days)	1.5 - 4.5	> 4.5 but the longer the better		
Total Selector F/M (kg BOD5/kg MLSS-days)	> 1.0	< 1.0 (the lower the better)	<= 1.0	Jenkins et al, 2004
Selector MCRT (days)	< 1.0	2.0 - 3.0+	1.0 - 2.0	Jenkins et al, 2004
No. of Selector Stages	1	2	3	Jenkins, 2004; Wanner, 1994
Aeration Basin DO (mg/L)	2.5 - 4.0	2.5 - 4.0 (air plants only)	> 1 - 2	Jenkins, 2004; Wanner, 1994
BOD/TSS Ratio	1.0 - 2.0	< 0.5 (the lower the better)		
ICZ F/M (kg BOD/kg TSS-days)	1.0 - 3.0	< 1.0 (the lower the better)	6	Wanner, 1994
Selector HRT w/o RAS (hours)	1.2 - 2.5	at least 1.2 but > 2.5 best		
Selector Vol/Total Basin Vol Ratio (%)	< 22.5	22.5 - 25.0	25	Wanner, 1994
Selector HRT w/ RAS (hours)	< 0.75	> 1.5	0.75 - 2.0	Jenkins et al, 2004
ICZ HRT w/ RAS (hours)	< 1.4	1.4 - 1.6		
ICZ HRT w/o RAS (hours)	< 2.4	2.4 - 2.7		
Effluent Temperature (degree Celsius)	20 - 25	20 - 25 (27 - 30+ worst)	< 28	Wanner, 1994; Richard et al, 1985
No. of Aeration Basin Stages		not significant		
Activated Sludge Influent BOD (mg/L)		not significant		
% RAS Flow (%)		not significant	<= 100	Wanner, 1994
Effluent pH		not significant		

Figure F-4. Main Results Window.

You can also click on a parameter name to display pop-up box that contains more information about recommended range(s) for that parameter (Figure F-5). All of the information in the pop-up boxes can also be displayed together by opening an additional information summary window labeled “discussionFrm: Form” (Figure F-5).

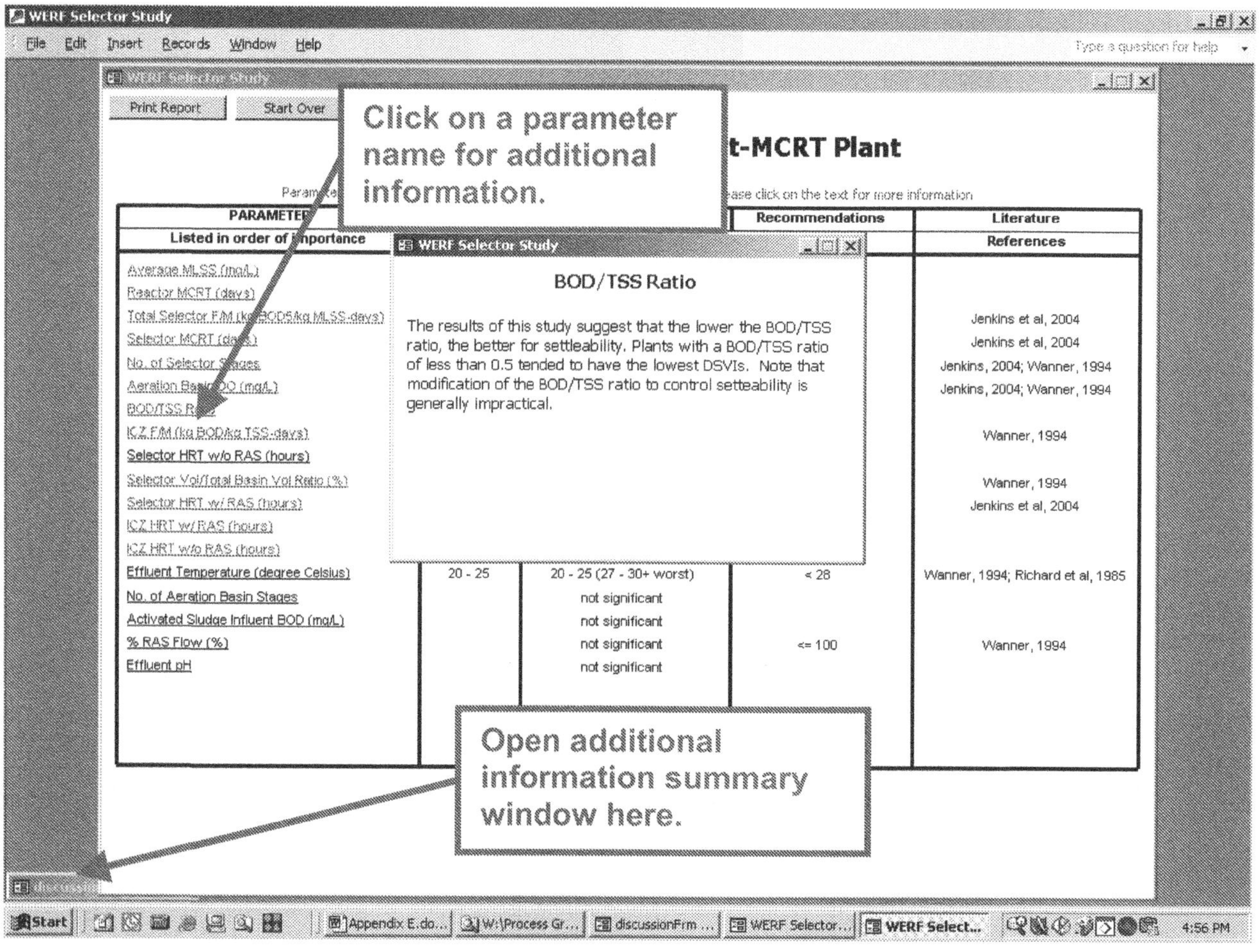

Figure F-5. Accessing Additional Information.

The additional information summary window is shown below (Figure F-6):

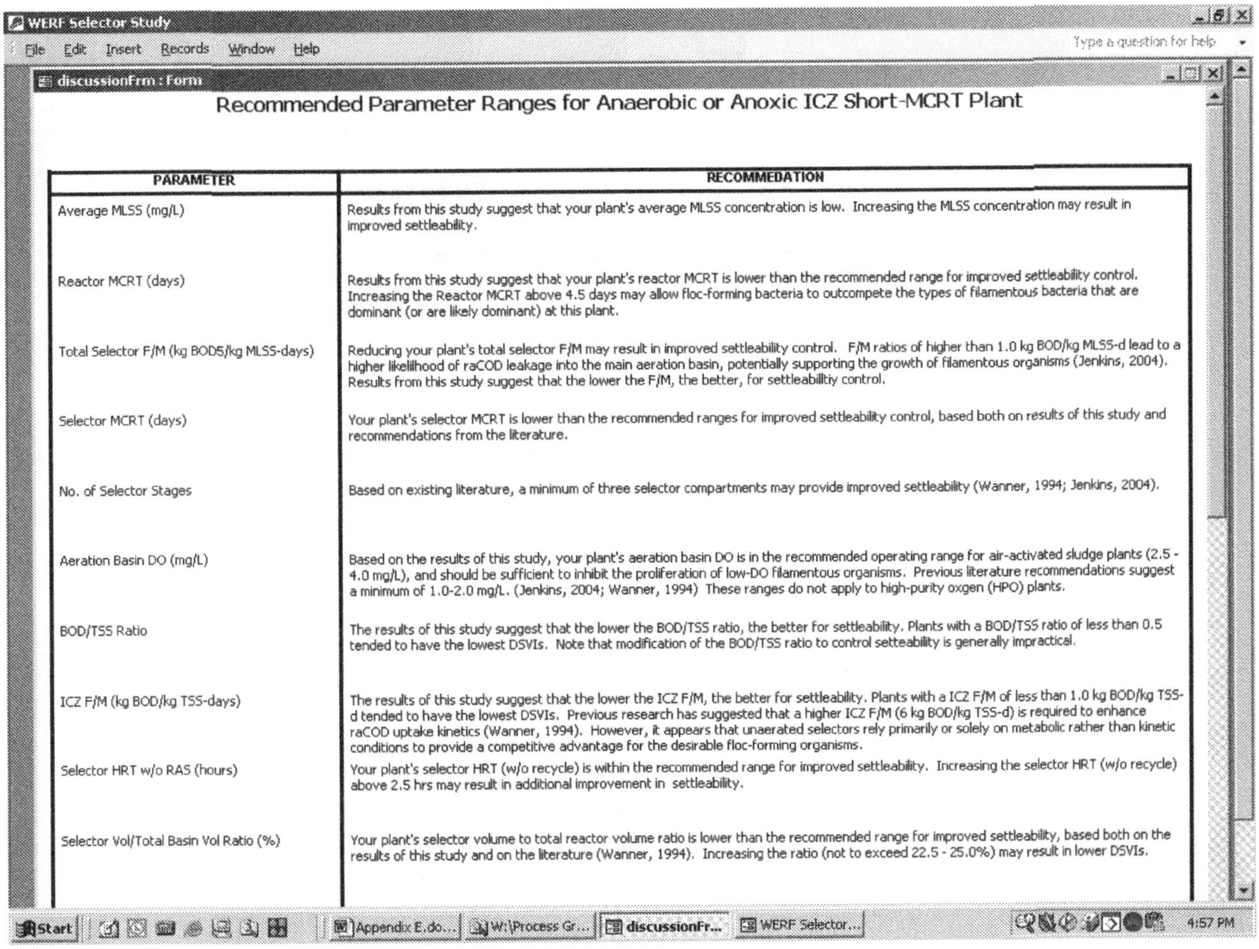

Recommended Parameter Ranges for Anaerobic or Anoxic ICZ Short-MCRT Plant

PARAMETER	RECOMMEDATION
Average MLSS (mg/L)	Results from this study suggest that your plant's average MLSS concentration is low. Increasing the MLSS concentration may result in improved settleability.
Reactor MCRT (days)	Results from this study suggest that your plant's reactor MCRT is lower than the recommended range for improved settleability control. Increasing the Reactor MCRT above 4.5 days may allow floc-forming bacteria to outcompete the types of filamentous bacteria that are dominant (or are likely dominant) at this plant.
Total Selector F/M (kg BOD5/kg MLSS-days)	Reducing your plant's total selector F/M may result in improved settleability control. F/M ratios of higher than 1.0 kg BOD/kg MLSS-d lead to a higher likelihood of raCOD leakage into the main aeration basin, potentially supporting the growth of filamentous organisms (Jenkins, 2004). Results from this study suggest that the lower the F/M, the better, for settleabiltiy control.
Selector MCRT (days)	Your plant's selector MCRT is lower than the recommended ranges for improved settleability control, based both on results of this study and recommendations from the literature.
No. of Selector Stages	Based on existing literature, a minimum of three selector compartments may provide improved settleability (Wanner, 1994; Jenkins, 2004).
Aeration Basin DO (mg/L)	Based on the results of this study, your plant's aeration basin DO is in the recommended operating range for air-activated sludge plants (2.5 - 4.0 mg/L), and should be sufficient to inhibit the proliferation of low-DO filamentous organisms. Previous literature recommendations suggest a minimum of 1.0-2.0 mg/L. (Jenkins, 2004; Wanner, 1994) These ranges do not apply to high-purity oxgen (HPO) plants.
BOD/TSS Ratio	The results of this study suggest that the lower the BOD/TSS ratio, the better for settleability. Plants with a BOD/TSS ratio of less than 0.5 tended to have the lowest DSVIs. Note that modification of the BOD/TSS ratio to control setteability is generally impractical.
ICZ F/M (kg BOD/kg TSS-days)	The results of this study suggest that the lower the ICZ F/M, the better for settleability. Plants with a ICZ F/M of less than 1.0 kg BOD/kg TSS-d tended to have the lowest DSVIs. Previous research has suggested that a higher ICZ F/M (6 kg BOD/kg TSS-d) is required to enhance raCOD uptake kinetics (Wanner, 1994). However, it appears that unaerated selectors rely primarily or solely on metabolic rather than kinetic conditions to provide a competitive advantage for the desirable floc-forming organisms.
Selector HRT w/o RAS (hours)	Your plant's selector HRT (w/o recycle) is within the recommended range for improved settleability. Increasing the selector HRT (w/o recycle) above 2.5 hrs may result in additional improvement in settleability.
Selector Vol/Total Basin Vol Ratio (%)	Your plant's selector volume to total reactor volume ratio is lower than the recommended range for improved settleabilty, based both on the results of this study and on the literature (Wanner, 1994). Increasing the ratio (not to exceed 22.5 - 25.0%) may result in lower DSVIs.

Figure F-6. Additional Information Summary Window.

Both the main results table and the additional information table can be printed using the "Print report" button on the main results window (Figure F-7).

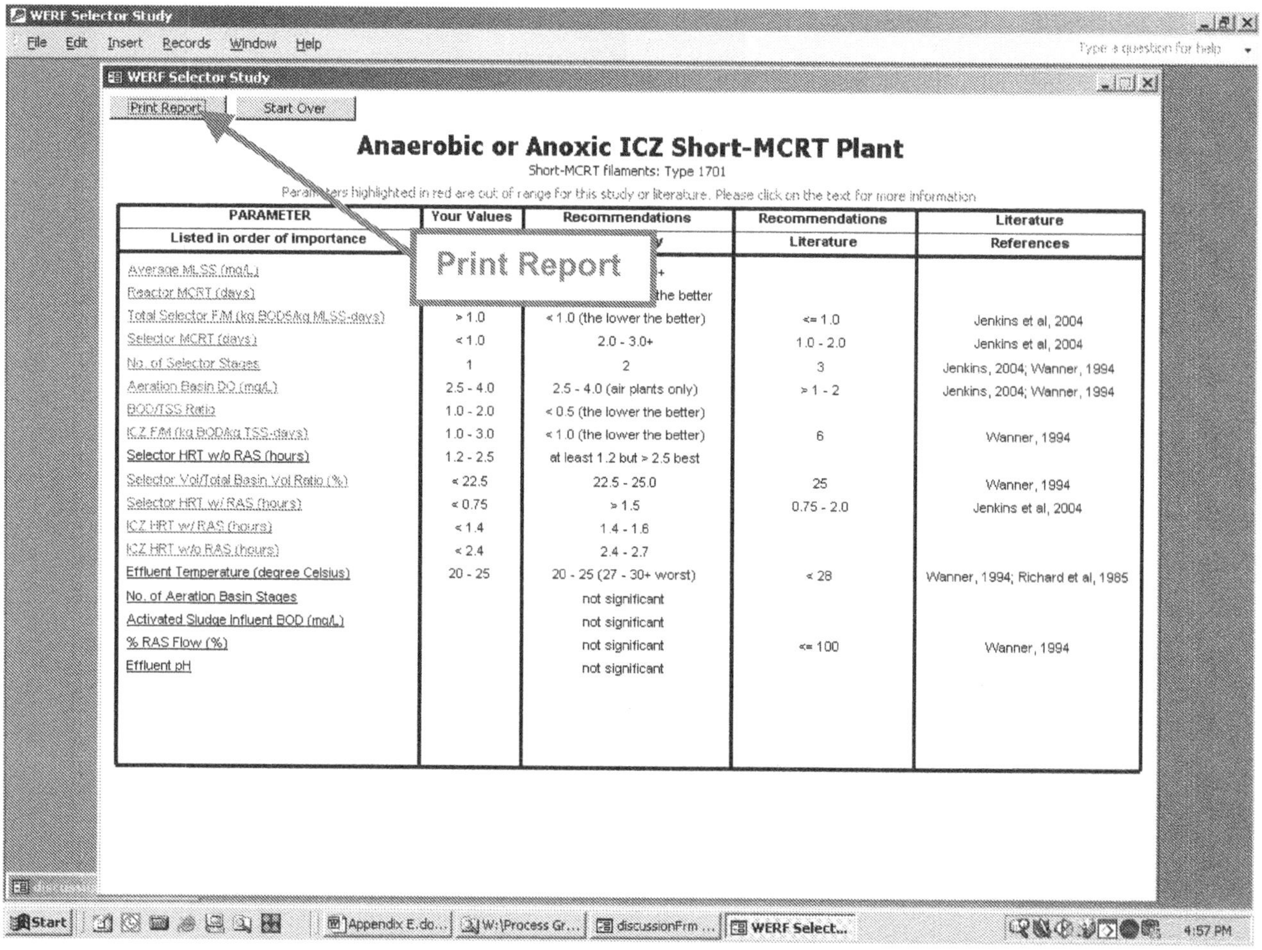

Figure F-7. Print Report Button.

Further Evaluation of Results

For more detailed information about the results, refer to the Regression Analysis section under Detailed Plant Investigations, Section 4.4.12. Regression analyses were performed separately for three categories of plants: short-MCRT plants with anoxic and anaerobic selectors, short-MCRT plants with aerobic selectors, and long-MCRT plants (all selector types). Results from the field study are presented for each category of selector, with a section summarizing data for each parameter correlated with sludge settleability.

REFERENCES

Albertson, O.E. 1987. The control of bulking sludges—from the early innovators to current practice. *Journal Water Pollution Control Federation*. 59(4):172-182.

Albertson, O.E. 1991. Bulking sludge control progress, practice and problems. *Water Science and Technology*. 23(4-6):835-846.

Albertson, O.E. 2005. Technology assessments: activated sludge bioselector processes—draft report submitted to Water Environment Research Foundation.

Albertson, O.E. and Hendricks, P. 1992. Bulking and foaming organism control at Phoenix, AZ WWTP. *Water Science and Technology*. 26(3-4):461-472.

Andreasen, K., Agertved, J., Petersen, J., and Skaarup, H. 1999. Improvement of sludge settleability in activated sludge plants treating effluent from pulp and paper industries. *Water Science and Technology*. 40(11-12):215-221.

Andreasen, K. and Nielsen, P.H. 1997. Application of microautoradiography to the study of substrate uptake by filamentous microorganisms in activated sludge. *Applied and Environmental Microbiology*. 63(9):3662-3668.

Andreasen, K. and Nielsen, P.H. 2000. Growth of *Microthrix parvicella* in nutrient removal activated sludge plants: studies of *in situ* physiology. *Water Research*. 34(5):559-1569.

Aruga, S., Kamagata, Y., Kohno, T., Hanada, S., Nakamura, K., and Kanagawa, T. 2002. Characterization of filamentous Eikelboom type 021N bacteria and description of *Thiothrix disciformis sp nov* and *Thiothrix flexilis sp nov*. *International Journal of Systematic and Evolutionary Microbiology*. 52(4):1309-1316.

Beccari, M., Dionisi, D., Giuliani, A., Majone, M., and Ramadori, R. 2002. Effect of different carbon sources on aerobic storage by activated sludge. *Water Science and Technology*. 45(6):157-168.

Beccari, M., Majone, M., Massanisso, P., and Ramadori, R. 1998. A bulking sludge with high storage response selected under intermittent feeding. *Water Research*. 32(11):3403-3413.

Beun, J.J., Paletta, F., Van Loosdrecht, M.C.M., and Heijnen, J.J. 2000. Stoichiometry and kinetics of poly-beta-hydroxybutyrate metabolism in aerobic, slow growing, activated sludge cultures. *Biotechnology and Bioengineering*. 67(4):379-389.

Blackall, L.L., Seviour, E.M., Cuninngham, M.A., Seviour, R.J., and Hugenholtz, P. 1994. '*Microthrix parvicella*' is a novel, deep branching member of the *Actinomycetes Subphylum*. *Systematic and Applied Microbiology*. 17(4):513-518.

Blackbeard, J.R., Gabb, D.M.D., Ekama, G.A., and Marais G.v.R. 1987. Identification of filamentous organisms in nutrient removal activated sludge plants in South Africa. *Water SA*. 14:29-34.

Bortone, G., Cech, J.S., Bianchi, R., and Tilche, A. 1995. Effects of an anaerobic zone in a textile wastewater treatment plant. *Water Science and Technology*. 32(9-10):130-140.

Carter J.L. and McKinney, R.E. 1973. Effects of iron on activated sludge treatment. *J. Environ. Engineer Div. Amer. Soc. Civil Eng.* 99:135.

Casey, T.G., Wentzel, M.C., and Ekama, G.A. 1999. Filamentous organism bulking in nutrient removal activated sludge systems. Paper 11: a biochemical/microbiological model for proliferation of Anoxic-Aerobic (AA) filamentous organisms. *Water SA*. 25(4):443-451.

Casey, T.G., Wentzel, M.C., Ekama, G.A., Loewenthal, R.E., and Marais, G.v.R. 1994. A hypothesis for the causes and control of Anoxic-Aerobic (AA) filament bulking in nutrient removal activated sludge systems. *Water Science and Technology*. 29(7):203-212.

Casey, T.G., Wentzel, M.C., Loewenthal, R.E., Ekama, G.A., and Marais, G.v.R. 1992. A hypothesis for the cause of low F/M filament bulking in nutrient removal activated sludge systems. *Water Research*. 26(6): 867-869.

Chambers, B. 1982. Effect of longitudinal mixing and anoxic zones on settleability of activated sludge. Chapter 10 in *Bulking of Activated Sludge: Preventative and Remedial Methods*. Chambers, B., and Tomlinson, E.J., eds. Ellis Horwood Ltd., Chichester, England.

Chambers, B. and Tomlinson, E.J. 1982. *Bulking of Activated Sludge: Preventative and Remedial Methods*, Ellis Horwood Ltd., West Sussex, England.

Chiesa, S.C. and Irvine, R.L. 1982. Growth and control of filamentous microbes in activated sludge—an integrated hypothesis. Presented at the 55th annual conference of the *Water Pollution Control Federation*, St. Louis, MO.

Chua, H. and Le, K.Y. 1994. A survey of filamentous foaming in an activated sludge plant in Hong Kong. *Water Science and Technology*. 30(11):251-254.

Chua, H., Tan, K.N., and Cheung, M.W.L. 1996. Filamentous growth in activated sludge. *App. Biochem. Biotech*. 57/58:851-856.

Chudoba, J., Cech, J.S., Farkac, J., and Grau, P. 1985. Control of activated sludge filamentous bulking—experimental verification of a kinetic selection theory. *Water Research*. 19(2):191-196.

Chudoba, J., Dohanyos, M., and Grau, P. 1982. Control of activated sludge filamentous bulking. 4. Effect of sludge regeneration. *Water Science and Technology*. 14(1-2):73-93.

Chudoba, J., Ottova, V., and Madera, V. 1973a. Control of activated sludge filamentous bulking. I. Effect of hydraulic regime of degree of mixing in an aeration tank. *Water Research.* 7:1163.

Chudoba, J., Grau, P., and Ottova, V. 1973b. Control of activated sludge filamentous bulking. II. Selection of microorganisms by means of a selector. *Water Research.* 7(10):1389-1406.

Chudoba, J. and Wanner, J. 1987. The control of bulking sludge: from the early innovators to current practice. *Journal Water Pollution Control Federation.* 59:172.

Clark, S.F., Krumsick, T.A., Bishop, R.P., Daigger, G.T., and Linder, C. 2001. Performance of a step-feed activated sludge system with anoxic selectors. *Proceedings, Water Environment Federation Annual Conference*, Alexandria, VA.

Contreras, E.M., Giannuzzi, L., and Zaritzky, N.E. 2000. Growth kinetics of the filamentous microorganism *Sphaerotilus natans* in a model system of a food industry wastewater. *Water Research.* 34(18):4455-4463.

Contreras, E.M., Giannuzzi, L., and Zaritzky, N.E. 2002. Competitive growth kinetics of *Sphaerotilus natans* and Acinetobacter anitratus. *Water Science and Technology.* 46(1-2):45-48.

Daigger, G.T. and Grady, C.P.L. 1982a. An assessment of the role of physiological adaptation in the transient-response of bacterial cultures. *Biotechnology and Bioengineering.* 24(6):1427-1444.

Daigger, G.T. and Grady, C.P.L. 1982b. The dynamics of microbial-growth on soluble substrates —a unifying theory. *Water Research.* 16(4):365-382.

Daigger, G.T. and Nicholson, G.A. 1990. Performance of 4 full-scale nitrifying waste-water treatment plants incorporating selectors. *Research Journal of the Water Pollution Control Federation.* 62(5):676-683.

Daigger, G.T., Robbins, M.A., Jr., and Marshall, B.R. 1985. The design of a selector to control low F/M filamentous bulking. *Journal Water Pollution Control Federation.* 57:2200.

Davidson, A.B. 1948. *Laboratory Test Work and Development of Patent Application for the Anaerobic and Aerobic Process.* Cincinnati, Ohio: Schenley Distillery.

Davoli, D., Madoni, P., Guglielmi, L., Pergetti, M., and Barilli, S. 2002. Testing the effect of selectors in the control of bulking and foaming in full-scale activated sludge plants. *Water Science and Technology.* 46(1-2):495-498.

DeLurgio, S.A. 1998. *Forecasting Principles and Applications.* Irwin/McGraw-Hill: New York.

Di Marzio, W.D. 2002. First results from a screening of filamentous organisms present in Buenos Aires' activated sludge plants. *Water Science and Technology.* 46(1):119-122.

Dick, R.I. and Vesilind, P.A. 1969. The sludge volume index—what is it? *Journal Water Pollution Control Federation.* 41(7), 1285-1291.

Dionisi, D., Levantesi, C., Renzi, V., Tandoi, V., and Majone, M. 2002. PHA storage from several substrates by different morphological types in an anoxic/aerobic SBR. *Water Science and Technology*. 46(1-2), 337-344.

Dionisi, D., Majone, M., Tandoi, V., and Beccari, M. 2001. Sequencing batch reactor: Influence of periodic operation on performance of activated sludges in biological wastewater treatment. *Industrial & Engineering Chemistry Research*. 40(23):5110-5119.

Donaldson, W. 1932. Use of activated sludge increasing. *Civil Engineering*. 2(3):167-169.

Donkin, M.J. 1999. Statistical analysis to associate specific microorganism populations with operating conditions and treatment performance in laboratory-scale systems treating dairy processing wastewater. *Environmental Technology*. 20(10):1085-1093.

Dueholm, T.E., Andreasen, K.H., and Nielsen, P.H. 2001. Transformation of lipids in activated sludge. *Water Science and Technology*. 43(1):165-172.

Duine, A. and Kunst, S. 2002. Control of bulking sludge caused by type 021N and type 0961 in an industrial wastewater treatment plant with an aerobic selector. *Water Science and Technology*. 46(1-2):29-33.

Eikelboom, D.H. 1975. Filamentous organisms observed in bulking activated sludge. *Water Res*. 9:365.

Eikelboom, D.H. 1977. Identification of filamentous organisms in bulking activated sludge. *Prog. Wat. Tech*. 8:153.

Eikelboom, D.H. 1994. The *Microthrix parvicella* puzzle. *Water Science and Technology*. 29 (7):271-279.

Eikelboom, D.H. 2000. *Process Control of Activated Sludge Plants by Microscopic Investigation*. IWA Publishing, London.

Ekama, G.A., Barnard, J.L., Gunthert, F.W., Krebs, P., McCorquodale, J.A., Parker, D.S., and Wahlberg, E.J. 1997. Secondary settling tanks: theory, modeling, design and operation. IAWQ, London. Chapter 3.

Ekama, G.A. and Marais, G.V. 1986. The implications of the IAWPRC hydrolysis hypothesis on low F/M bulking. *Water Science and Technology*. 18(6):11-19.

Fainsod, A., Pagilla, K.R., Jenkins, D., Pitt, P.A., and Mamais, D. 1999. The effect of anaerobic selectors on nocardioform organism growth in activated sludge. *Water Environmental Research* 71(6):1151-1157.

Gabb, D.M.D. 1988. Filamentous bulking in long mean cell residence time activated sludge processes. Ph.D. dissertation, Dept. of Civil Eng., University of California, Berkeley, CA.

Gabb, D.M.D., Ekama, G.A., Jenkins, D., and Marais, G.v.R. 1989. The incidence of *Sphaerotilus natans* in laboratory scale activated sludge systems. *Water Science Technology,* 21:29-41.

Gabb, D.M.D. and Jenkins, D. 1991a. Influence of substrate type on the ability of selectors to control filamentous bulking in activated sludge. Presented at the 64th Annual Conference of the Water Pollution Control Federation (now WEF), Toronto, Ontario, Canada.

Gabb, D.M.D., Still, A., Ekama, G.A., Jenkins, D., and Marais, G.v.R. 1991b. The selector effect on filamentous bulking in long sludge age activated sludge systems. *Water Science and Technology*. 23:867-877.

Grady, C.P.L., Jr., Daigger, G.T., and Lim, H.C. 1999. *Biological Wastewater Treatment, 2nd ed.,* Marcel Dekker, New York, 109.

Grau, P., Chudoba, J., and Dohanyos, M. 1982. Theory and practice of accumulation-regeneration approach to the control of activated sludge filamentous bulking. Chapter 7 in *Bulking of Activated Sludge: Preventative and Remedial Methods*. Chambers, B., and Tomlinson, E. J., eds. Ellis Horwood Ltd., Chichester, England. 111-127.

Guida, M., Cesaro, G., Lipardi, I.L., and Melluso, G. 2002. A full scale application in the control of the filamentous bulking generated by Type 021NF. *Thiothrix sp. Water Science and Technology*. 46(1-2):507-510.

Gujer, W., Henze, M., Mino, T., and van Loosdrecht, M. 1999. Activated sludge model No. 3. *Water Science and Technology*. 39(1):183-193.

Hagland, E., Westlund, A.D., and Rothman, M. 1998. Sludge retention time in the secondary clarifier effects on the growth of *Microthrix parvicella. Water Science and Technology*. 37(4-5):47-50.

Hao, O.J. 1982. Isolation, characterization, and continuous culture kinetics of a new *Sphaerotilus* species involved in low oxygen activated sludge bulking. Ph.D. dissertation, Dept. of Civil Eng., University of California, Berkeley, CA.

Harleman, D.R.F. 1964. The significance of longitudinal dispersion in the analysis of pollution in estuaries. *Proceedings 2d International Conference on Water Pollution Research,* Tokyo, Pergamon New York. *Wastewater Engineering: Treatment, Disposal, Reuse, 2nd ed.*, Metcalf and Eddy. McGraw-Hill Inc.

Henze, M., Gujer, W., Mino, T., Matsuo, T., Wentzel, M.C., and Marais, G.v.R. 1995. *Activated sludge model No.2. IAWQ Scientific and Technical Report, No.3*. London: IAWQ.

Holmstrom, H., Bosander, J., Dahlberg, A.G., DillnerWestlund, A., Flyborg, L., and Jokinen, K. 1996. Severe bulking and foaming at the Himmerfjarden WWTP. *Water Science and Technology*. 33(12):127-135.

Houtmeyers, J., van den Eynde, E., Poffe, R., and Verachteret, H. 1980. Relations between substrate feeding pattern and development of filamentous bacteria in activated sludge processes. Part 1: Influence of process parameters. *European J. Appl. Microbio. Biotechnol.* 9:63.

Howarth, R., Unz, R.F., Seviour, E.M., Seviour, R.J., Blackall, L.L., Pickup, R.W., Jones, J.G., Yaguchi, J., and Head, I.M. 1999. Phylogenetic relationships of filamentous sulfur bacteria (*Thiothrix spp.* and Eikelboom Type 021N bacteria) isolated from wastewater-treatment plants and description of *Thiothrix eikelboomii sp nov., Thiothrix unzii sp nov., Thiothrix fructosivorans sp nov* and *Thiothrix defluvii sp nov. International Journal of Systematic Bacteriology.* 49:1817-1827.

Hwu, C.S., Tseng, S.K., Yuan, C.Y., Kulik, Z., and Lettinga, G. 1998. Biosorption of long-chain fatty acids in UASB treatment process. *Water Research.* 32(5):1571-1579.

Jenkins, D., Richard, M.G., and Daigger, G.T. 1984. *Manual on the Causes and Control of Activated Sludge Bulking, Foaming. Water Research Commission.* Pretoria, South Africa.

Jenkins, D., Richard, M.G., and Daigger, G.T. 1993. *Manual on the Causes and Control of Activated Sludge Bulking, Foaming, and Other Solids Separation Problems* (2nd ed.). Michigan: Lewis Publishers.

Jenkins, D., Richard, M.G., and Daigger, G.T. 2004. *Manual on the Causes and Control of Activated Sludge Bulking, Foaming, and Other Solids Separation Problems* (3rd ed.). Boca Raton, Fla.: CRC Press, Lewis Publishers.

Kaewpipat, K. and Grady, C.P.L. 2002. Microbial population dynamics in laboratory-scale activated sludge reactors. *Water Science and Technology.* 46(1-2):19-27.

Kampfer, P., Weltin, D., Hoffmeister, D., and Dott, W. 1995. Growth requirements of filamentous bacteria isolated from bulking and scumming sludge. *Water Research.* 29(6):1585-1588.

Kappeler, J. and Gujer, W. 1993. Development of a mathematical model for “aerobic bulking”. *Water Research.* 28(2):303-310.

Kappeler, J. and Gujer, W. 1994. Influences of waste-water composition and operating-conditions on activated-sludge bulking and scum formation. *Water Science and Technology.* 30(11):181-189.

Keller, G. and Warrack, B. 2000. *Statistics for Management and Economics.* Duxbury: CA.

Knoop, S. and Kunst, S. 1998. Influence of temperature and sludge loading on activated sludge settling, especially on *Microthrix parvicella. Water Science and Technology.* 37(4-5):27-35.

Kohno, T. 1989. Morphology, physiology, and nutrition of a sulphur-oxidizing filamentous organism isolated from activated sludge. *Water Science and Technology.* 20(11-12):241-247.

Kohno, T., Sei, K., and Mori, K. 2002. Characterization of type 1851 organism isolated from activated sludge samples. *Water Science and Technology*. 46(1-2):111-114.

Krishna, C. and Van Loosdrecht, M.C.M. 1999. Effect of temperature on storage polymers and settleability of activated sludge. *Water Research*. 33(10):2374-2382.

Kruit, J., Hulsbeek, J., and Visser, A. 2002. Bulking sludge solved? *Water Science and Technology*. 46(1-2):457-464.

Larkin, J.M. and Shinabarger, D.L. 1983. Characterization of *Thiothrix nivea*. *International Journal of Systematic Bacteriology*. 33(4):841-846.

Lebek, M. and Rosenwinkel, K.H. 2002. Control of the growth of *Microthrix parvicella* by using an aerobic selector—results of pilot and full scale plant operation. *Water Science and Technology*. 46(1-2):491-494.

Lee, S.-E., Koopman, B.L., Bode, H., and Jenkins, D. 1983. Evaluation of alternative sludge settleabilitly indices. *Water Research*. 17(10):1421-1426.

Lee, S.-E., Koopman, B.L., Jenkins, D., and Lewis, R.F. 1982. The effect of aeration basin configuration on activated sludge bulking at low organic loading. *Water Science and Technology*. 14:407.

Liao, J.Y., Lou, I.C., and los Reyes, F.L. 2004. Relationship of species-specific filament levels to filamentous bulking in activated sludge. *Applied and Environmental Microbiology*. 70(4):2420-2428.

Madoni, P. and Davoli, D. 1997. Testing the control of filamentous microorganisms responsible for foaming in a full-scale activated-sludge plant running with initial aerobic or anoxic contact zones. *Bioresource Technology*. 60(1):43-49.

Majone, M., Massanisso, P., Carucci, A., Lindrea, K., and Tandoi, V. 1996. Influence of storage on kinetic selection to control aerobic filamentous bulking. *Water Science and Technology*. 34(5-6):223-232.

Mamais, D., Andreadakis, A., Noutsopoulos, C., and Kalergis, C. 1998. Causes of, and control strategies for, *Microthrix parvicella* bulking and foaming in nutrient removal activated sludge systems. *Water Science and Technology*. 37(4-5):9-17.

Marshall, R. and Richard, M. 2000. Selectors in pulp and paper mill-activated sludge operations: do they work? *Pulp and Paper Canada*. 101(3):48-53.

Marten, W.L. and Daigger, G.T. 1997. Full-scale evaluation of factors affecting the performance of anoxic selectors. *Water Environmental Research*. 69(7):1272-1281.

Martins, A.M.P., Heijnen, J.J., and van Loosdrecht, M.C.M. 2003a. Effect of feeding pattern and storage on the sludge settleability under aerobic conditions. *Water Research*. 37(11):2555-2570.

Martins, A.M.P., Heijnen, J.J., and van Loosdrecht, M.C.M. 2003b. Effect of dissolved oxygen concentration on sludge settleability. *Applied Microbiology and Biotechnology*. 62(5-6):586-593.

Martins, A.M.P., Pagilla, K., Heijnen, J.J., and van Loosdrecht, M.C.M. 2004a. Filamentous bulking sludge—a critical review. *Water Research*. 38(4):793-817.

Martins, A.M.P., Heijnen, J.J., and van Loosdrecht, M.C.M. 2004b. Bulking sludge in biological nutrient removal systems. *Biotechnology and Bioengineering*. 86(2):125-135.

Merkel, W. 1971. Untersuchungen uber das verhalten des belebten schlammes im system belebungsbecken-nachklarbecken. *Gewasserschutz, Wasser-Abwasser* (ed. B. Bohnke), vol.5, D82, Institut fur Siedlungswasserwirtschaft, Technischen Hochschule Aachen, Aachen. Cited by Ekama et. al. (1997). p. 42.

Mino, T. 1995. Survey of filamentous microorganisms in activated sludge processes in Bangkok, Thailand. *Water Science and Technology*. 31(9):193-202.

Monod, J. 1949. The growth of bacteria cultures. *Annual Review of Microbiology*. 3:371-394.

Montgomery, D.C. and Peck, E.A. 1982. Introduction to linear regression analysis. *Minitab Statistical Software Users Guide (2000)*. John Wiley & Sons.

Morgan-Sagastume, F. and Allen, D.G. 2003. Effects of temperature transient conditions on aerobic biological treatment of wastewater. *Water Research*. 37(15):3590-3601.

Nielsen, J.L., Christensen, D., Kloppenborg, M., and Nielsen, P.H. 2003. Quantification of cell-specific substrate uptake by probe-defined bacteria under *in situ* conditions by microautoradiography and fluorescence *in situ* hybridization. *Environmental Microbiology*. 5(3):202-211.

Nielsen, P.H., de Muro, M.A., and Nielsen, J.L. 2000. Studies on the *in situ* physiology of *Thiothrix spp.* present in activated sludge. *Environmental Microbiology*. 2(4):389-398.

Nielsen, P.H., Roslev, P., Dueholm, T.E., and Nielsen, J.L. 2002. *Microthrix parvicella*, a specialized lipid consumer in anaerobic-aerobic activated sludge plants. *Water Science and Technology*. 46(1-2):73-80.

Nikolavcic, B. and Svardal, K. 2000. Biological treatment of potato-starch wastewater design and application of an aerobic selector. *Water Science and Technology*. 41(9):251-258.

Noutsopoulos, C., Mamais, D., and Andreadakis, A.D. 2002. The effect of reactor configuration and operational mode on *Microthrix parvicella* bulking and foaming in nutrient removal activated sludge systems. *Water Science and Technology*. 46(1-2):61-64.

Odintsova, E.V. and Dubinina, G.A. 1991. Developmental cycle, reproduction, and ultrastructure of *Thiothrix-Ramosa. Microbiology*. 60(2):214-219.

Odintsova, E.V., Wood, A.P., and Kelly, D.P. 1993. Chemolithoautotrophic growth of *Thiothrix-Ramosa. Archives of Microbiology*. 160(2):152-157.

Osborn, D.W., Lotter, L.H., Pitman, A.R., and Nicholls, H.A. 1986. Operational and design aspects. Chapter 8 in *Enhancement of Biological Phosphate Removal by Altering Process Feed Composition*. Report to the Water Research Commission, RSA, No. 137/1/86.

Palm, J.C., Jenkins, D., and Parker, D.S. 1980. Relationship between organic loading, dissolved oxygen concentration and sludge settleability in the completely-mixed activated sludge process. *Journal Water Pollution Control Federation*. 52(10):2484-2506.

Parker, D.S., Geary, S., Jones, G., McIntyre, L., Oppenheim, S., Pedregon, V., Pope, R., Richard, T., Voight, C., Volpe, G., Willis, J., and Witzgall, R. 2003. Making classifying selectors work for foam elimination in the activated sludge process. *Water Environ. Res*. 75:83.

Pasveer, A. 1969. A case of filamentous activated sludge. *Journal Water Pollution Control Federation*. 41(7):1340-52.

Pellegrin,V., Juretschko, S., Wagner, M., and Cottenceau, G. 1999. Morphological and biochemical properties of a *Sphaerotilus sp.* isolated from paper mill slimes. *Applied and Environmental Microbiology*. 65(1):156-162.

Pernelle, J.J., Gaval, G., Cotteux, E., and Duchene, P. 2001. Influence of transient substrate overloads on the proliferation of filamentous bacterial populations in an activated sludge pilot plant. *Water Research* 35(1)129-34.

Pipes, W.O. 1967. Bulking of activated sludge. *Advanced Application of Microbiology*. 9:185-234.

Prendl, L. and Kroiss, H. 1998. Bulking sludge prevention by an aerobic selector. *Water Science and Technology*. 38(8-9):19-27.

Quemeneur, M. and Marty, Y. 1994. Fatty-Acids and sterols in domestic wastewaters. *Water Research*. 28(5):1217-1226.

Raunkjaer, K., Hvitvedjacobsen, T., and Nielsen, P.H. 1994. Measurement of pools of protein, carbohydrate and lipid in domestic waste-water. *Water Research*. 28(2):251-262.

Rensink, J.H. and Donker, H.J.G.W. 1991. The Effect of Contact Tank Operation on Bulking-Sludge and Biosorption Processes. *Water Science and Technology*. 23(4-6):857-866.

Rensink, J.H., Donker, H.J.G.W., and Ijwema, T.S.J. 1982. The influence of feed pattern on sludge bulking. Chapter 9 in *Bulking of Activated Sludge: Preventative and Remedial Methods*. Chambers, B., and Tomlinson, E. J., eds. Ellis Horwood Ltd., Chichester, England.

Richard, M.G., Hao, O., and Jenkins, D. 1982. Growth kinetics of *Sphaerotilus* species and their significance in activated sludge bulking. Presented at the 55th Annual Conference of the Water Pollution Control Federation, St. Louis, MO.

Richard, M.G., Shimizu, G.P., and Jenkins, D. 1985. The growth physiology of the filamentous organism Type 021N and its significance to activated-sludge bulking. *Journal Water Pollution Control Federation*. 57(12):1152-1162.

Richard, M.G., Shimizu, G., and Jenkins, D. 1984. The growth physiology of the filamentous organism Type 021N and its significance to activated sludge bulking. Presented at the 57th Annual Conference of the Water Pollution Control Federation, New Orleans, LA.

Richard, M.G., Shimizu, G., Jenkins, D., Williams, T., and Unz, R.F. 1983. Isolation and characterization of *Thiothrix* and *Thiothrix*-like filamentous organisms from bulking activated sludge. Presented at the annual meeting of the American. Society. for Microbiology, New Orleans, La. (Abs. Ann. Meet., Q60, 270).

Rossetti, S., Blackall, L., Levantesi, C., Uccelletti, D., and Tandoi, V. 2003. Phylogenetic and physiological characterization of a heterotrophic, chemolithoautotrophic *Thiothrix* strain isolated from activated sludge. *International Journal of Systematic and Evolutionary Microbiology*. 53:1271-1276.

Rothman, M. 1998. Operation with biological nutrient removal with stable nitrification and control of filamentous growth. *Water Science and Technology*. 37(4-5):549-554.

Salehizadeh, H. and van Loosdrecht, M.C.M. 2004. Production of polyhydroxyalkanoates by mixed culture: recent trends and biotechnological importance. *Biotechnology Advances*. 22(3):261-279.

Satoh, H., Iwamoto, Y., Mino, T., and Matsuo, T. 1998. Activated sludge as a possible source of biodegradable plastic. *Water Science and Technology*. 38(2):103-109.

Schuler, A.J., Jenkins, D., and Ronen, P. 2001. Microbial storage products, biomass density, and settling properties of enhanced biological phosphorus removal activated sludge. *Water Science and Technology*. 43(1):173-180.

Shao, Y.J. 1983. The mechanism and design of anoxic selectors for the control of low F/M filamentous bulking. Ph.D. dissertation, Department of Civil Engineering, University of California, Berkeley, CA.

Shao,Y.J. and Jenkins. D. 1989. The use of anoxic selectors for the control of low F/M activated-sludge bulking. *Water Science and Technology*. 21(6-7):609-619.

Slijkhuis, H. 1983. *Microthrix parvicella*, a filamentous bacterium isolated from activated-sludge—cultivation in a chemically defined medium. *Applied and Environmental Microbiology*. 46(4):832-839.

Slijkhuis, H. 1983. The physiology of the filamentous bacterium *Microthrix parvicella*. Ph. D thesis, Wageningen, Holland.

Slijkhuis, H. and Deinema, M.H. 1988. Effect of environmental-conditions on the occurrence of *Microthrix-Parvicella* in activated-sludge. *Water Research*. 22(7):825-828.

Smolders, G.J.F., Vandermeij, J., Vanloosdrecht, M.C.M., and Heijnen, J.J. 1995. A structured metabolic model for anaerobic and aerobic stoichiometry and kinetics of the biological phosphorus removal process. *Biotechnology and Bioengineering*. 47(3):277-287.

Soddell, J.A. and Seviour, R.J. 1995. Relationship between temperature and growth of organisms causing *Nocardia* foams in activated-sludge plants. *Water Research*. 29(6):1555-1558.

Still, D.A., Ekama, G.A., Wentzel, M.C., Casey, T.G., and Marais, G.v.R. 1996. Filamentous organism bulking in nutrient removal activated sludge systems. Paper 2: Stimulation of the selector effect under aerobic conditions. *Water SA*. 22(2):97-118.

Still, D., Blackbeard, J.R., Ekama, G.A., and Marais, G.v.R. 1986. The effect of feeding patterns on sludge growth rate and sludge settleability. Res. Rept. No. W55, Department of Civil Engineering, University of Cape Town.

Stobbe, C.T. 1964. Uber das verhalten des belebten schlammes in aufsteigender wasserbewegung. Veroffentilichungen des institutes fur siedlungswasserwirtschaft der technischen hochschule hannover, vol. 18. Hannover. Cited by Ekama et. al. (1997) p. 42.

Strom, P.F. and Jenkins, D. 1984. Identification and significance of filamentous microorganisms in activated-sludge. *Journal Water Pollution Control Federation*. 56(5):449-459.

Tandoi, V., Caravaglio, N., Balsamo, D.D., Majone, M., and Tomei, M.C. 1994. Isolation and physiological characterization of *Thiothrix Sp. Water Science and Technology*. 29(7):261-269.

Tandoi, V., Rossetti, S., Blackall, L.L., and Majone, M. 1998. Some physiological properties of an Italian isolate of "*Microthrix parvicella*". *Water Science and Technology*. 37(4-5):1-8.

Tchobanoglous, G., Burton, F.L., and Stensel, H.D. 2003. *Wastewater Engineering Treatment and Reuse*. McGraw Hill.

Tomlinson, E.J. 1982. The emergence of the bulking problem and the current situation in the UK. In *Bulking of Activated Sludge: Preventive and Remedial Methods*, eds. B. Chambers and J. Tomlinson. Chap. 1: 17-23. Ellis Horwood Ltd., Chichester, England..

Tsai, C-H. 2001. *Forecasting and Managerial Research Methods*. CEO Press.

Tsai, M.W., Wentzel, M.C., and Ekama, G.A. 2003. The effect of residual ammonia concentration under aerobic conditions on the growth of *Microthrix parvicella* in biological nutrient removal plants. *Water Research*. 37(12):3009-3015.

Tsezos, M. and Bell, J.P. 1989. Comparison of the biosorption and desorption of hazardous organic pollutants by live and dead biomass. *Water Research.* 23(5):561-568.

van den Eynde, E., Geerts, J., Maes, B., and Verachtert, H. 1983. Influence of the feeding pattern on the glucose-metabolism of *Arthrobacter sp* and *Sphaerotilus natans*, growing in chemostat culture, simulating activated-sludge bulking. *European Journal of Applied Microbiology and Biotechnology*. 17(1):35-43.

van den Eynde, E., Vriens, L., and Verachtert, H. 1982a. Relationship between substrate feeding pattern and development of filamentous bacteria in activated sludge processes. III. Application with industrial wastewaters. *European Journal of Applied Microbiology and Biotechnology*. 15:246.

van den Eynde, E., Houtmeyers, J., and Verachtert, H. 1982b. Relationship between substrate feeding pattern and development of filamentous bacteria in activated sludge. Chapter 8 in *Bulking of Activated Sludge: Preventative and Remedial Methods*. Chambers, B., and Tomlinson, E. J., eds. Ellis Horwood Ltd., Chichester, England.

Van Niekerk, A.M. 1985. Competitive growth of flocculant and filamentous microorganisms in activated sludge systems. Ph.D. dissertation, Department of Civil Engineering, University of California, Berkeley, Calif.

Van Niekerk, A.M., Jenkins, D., and Richard, M.G. 1987. The competitive growth of *Zoogloea ramigera* and Type-021N in activated-sludge and pure culture—a model for low F-M bulking. *Journal Water Pollution Control Federation*. 59(5):262-273.

Verachtert, H., van den Eynde, E., Poffe, R., and Houtmeyers, J. 1980. Relationship between substrate feeding pattern and development of filamentous bacteria in activated sludge processes. II. Influence of substrate present in influent. *European Journal of Applied Microbiology and Biotechnology*, 9:137.

Wagner, M., Amann, R., Kampfer, P., Assmus, B., Hartmann, A., Hutzler, P., Springer, N., and Schleifer, K.H. 1994. Identification and *in situ* detection of gram-negative filamentous bacteria in activated-sludge. *Systematic and Applied Microbiology*. 17(3):405-417.

Wakefield, R.W. and Slim, J.A. 1987. The practical application of various techniques to control sludge bulking. Presented at the biennial conference and exhibition of the IAWPRC (S.A. branch), Port Elizabeth, RSA.

Wakelin, N.G. and Forster, C.F. 1997. An investigation into microbial removal of fats, oils and greases. *Bioresource Technology*. 59(1):37-43.

Wanner, J. 1994. *Activated Sludge Bulking and Foaming Control*. Lancaster, PA: Technomic Publishing.

Wanner, J. and Grau, P. 1988. Filamentous bulking in nutrient removal activated-sludge systems. *Water Science and Technology*. 20(4-5):1-8.

Wheeler, M., Jenkins, D., and Richard, M.G. 1983. The use of selectors for bulking control at the Hamilton, Ohio, USA, Water Pollution Control Facility. Proceedings, workshop on design, and operation of large wastewater treatment plants. International Association for Water Pollution Research and Control, Vienna, Austria.

Williams, T.M. and Unz, R.F. 1985. Filamentous sulfur bacteria of activated-sludge—characterization of *Thiothrix, Beggiatoa,* and Eikelboom Type 021N strains. *Applied and Environmental Microbiology*. 49(4):887-898.

Wu, Y.C., Hsieh, H.N., Carey, D.F., and Ou, K.C. 1983. Control of activated sludge bulking. *J. Environ. Engineer Div. Amer. Soc. Civil Eng.* 110:472.

WASTEWATER UTILITY

Alabama
Montgomery Water Works & Sanitary Sewer Board

Alaska
Anchorage Water & Wastewater Utility

Arizona
Glendale, City of, Utilities Department
Mesa, City of
Peoria, City of
Phoenix Water Services Department
Pima County Wastewater Management
Safford, City of

Arkansas
Little Rock Wastewater Utility

California
Calaveras County Water District
Central Contra Costa Sanitary District
Corona, City of
Crestline Sanitation District
Delta Diablo Sanitation District
Dublin San Ramon Services District
East Bay Dischargers Authority
East Bay Municipal Utility District
Eastern Municipal Water District
El Dorado Irrigation District
Fairfield-Suisun Sewer District
Fresno Department of Public Utilities
Inland Empire Utilities Agency
Irvine Ranch Water District
Las Virgenes Municipal Water District
Livermore, City of
Los Angeles, City of
Los Angeles County, Sanitation Districts of
Napa Sanitation District
Orange County Sanitation District
Palo Alto, City of
Riverside, City of
Sacramento Regional County Sanitation District
San Diego Metropolitan Wastewater Department, City of
San Francisco, City & County of
San Jose, City of
Santa Barbara, City of
Santa Cruz, City of
Santa Rosa, City of
South Bayside System Authority
South Coast Water District
South Orange County Wastewater Authority
Stege Sanitary District
Sunnyvale, City of
Union Sanitary District
West Valley Sanitation District

Colorado
Aurora, City of
Boulder, City of
Greeley, City of
Littleton/Englewood Water Pollution Control Plant
Metro Wastewater Reclamation District, Denver

Connecticut
Greater New Haven WPCA

District of Columbia
District of Columbia Water & Sewer Authority

Florida
Broward, County of
Fort Lauderdale, City of
Miami-Dade Water & Sewer Authority
Orange County Utilities Department
Reedy Creek Improvement District
Seminole County Environmental Services
St. Petersburg, City of
Stuart Public Utilities
Tallahassee, City of
Tampa, City of
Toho Water Authority
West Palm Beach, City of

Georgia
Atlanta Department of Watershed Management
Augusta, City of
Clayton County Water Authority
Cobb County Water System
Columbus Water Works
Fulton County
Gwinnett County Department of Public Utilities
Savannah, City of

Hawaii
Honolulu, City & County of

Idaho
Boise, City of

Illinois
American Bottoms Wastewater Treatment Plant
Greater Peoria Sanitary District
Kankakee River Metropolitan Agency
Metropolitan Water Reclamation District of Greater Chicago
Wheaton Sanitary District

Iowa
Ames, City of
Cedar Rapids Wastewater Facility
Des Moines, City of
Iowa City

Kansas
Johnson County Unified Wastewater Districts
Unified Government of Wyandotte County/ Kansas City, City of

Kentucky
Louisville & Jefferson County Metropolitan Sewer District

Louisiana
Sewerage & Water Board of New Orleans

Maine
Bangor, City of
Portland Water District

Maryland
Anne Arundel County Bureau of Utility Operations
Howard County Department of Public Works
Washington Suburban Sanitary Commission

Massachusetts
Boston Water & Sewer Commission
Upper Blackstone Water Pollution Abatement District

Michigan
Ann Arbor, City of
Detroit, City of
Holland Board of Public Works
Saginaw, City of
Wayne County Department of Environment
Wyoming, City of

Minnesota
Rochester, City of
Western Lake Superior Sanitary District

Missouri
Independence, City of
Kansas City Missouri Water Services Department
Little Blue Valley Sewer District
Metropolitan St. Louis Sewer District

Nebraska
Lincoln Wastewater System

Nevada
Henderson, City of
Reno, City of

New Jersey
Bergen County Utilities Authority
Ocean, County of
Passaic Valley Sewerage Commissioners

New York
New York City Department of Environmental Protection

North Carolina
Charlotte/Mecklenburg Utilities
Durham, City of
Metropolitan Sewerage District of Buncombe County
Orange Water & Sewer Authority

Ohio
Akron, City of
Butler County Department of Environmental Services
Columbus, City of
Metropolitan Sewer District of Greater Cincinnati
Northeast Ohio Regional Sewer District
Summit, County of

Oklahoma
Oklahoma City Water & Wastewater Utility Department
Tulsa, City of

Oregon
Clean Water Services
Eugene/Springfield Water Pollution Control
Gresham, City of
Water Environment Services

Pennsylvania
Philadelphia, City of
University Area Joint Authority

South Carolina
Charleston Commissioners of Public Works
Mount Pleasant Waterworks & Sewer Commission
Spartanburg Sanitary Sewer District

Tennessee
Cleveland, City of
Knoxville Utilities Board
Murfreesboro Water & Sewer Department
Nashville Metro Water Services

Texas
Austin, City of
Dallas Water Utilities
Denton, City of
El Paso Water Utilities
Fort Worth, City of
Gulf Coast Waste Disposal Authority
Houston, City of
San Antonio Water System
Trinity River Authority

Utah
Salt Lake City Corporation

Virginia
Alexandria Sanitation Authority

WERF SUBSCRIBERS

Arlington, County of
Fairfax County
Hampton Roads Sanitation District
Henrico, County of
Hopewell Regional Wastewater Treatment Facility
Loudoun County Sanitation Authority
Lynchburg Regional WWTP
Prince William County Service Authority
Richmond, City of
Rivanna Water & Sewer Authority

Washington
Everett, City of
King County Department of Natural Resources
Seattle Public Utilities
Sunnyside, Port of
Yakima, City of

Wisconsin
Green Bay Metro Sewerage District
Kenosha Water Utility
Madison Metropolitan Sewerage District
Milwaukee Metropolitan Sewerage District
Racine, City of
Sheboygan Regional Wastewater Treatment
Wausau Water Works

Australia
South Australian Water Corporation
Sydney Water Corporation
Water Corporation of Western Australia

Canada
Greater Vancouver Regional District
Regina, City of, Saskatchewan
Toronto, City of, Ontario
Winnipeg, City of, Manitoba

New Zealand
Watercare Services Limited

United Kingdom
Yorkshire Water Services Limited

STORMWATER UTILITY

California
Los Angeles, City of, Department of Public Works
Monterey, City of
Sacramento, County of
San Francisco, City & County of
Santa Rosa, City of
Sunnyvale, City of

Colorado
Aurora, City of
Boulder, City of

Georgia
Griffin, City of

Iowa
Cedar Rapids Wastewater Facility
Des Moines, City of

Kansas
Overland Park, City of

Maine
Portland Water District

Minnesota
Western Lake Superior Sanitary District

North Carolina
Charlotte, City of, Stormwater Services

Pennsylvania
Philadelphia, City of

Tennessee
Chattanooga Stormwater Management

Texas
Harris County Flood Control District, Texas

Washington
Bellevue Utilities Department
Seattle Public Utilities

STATE

Arkansas Department of Environmental Quality
Connecticut Department of Environmental Protection
Fresno Metropolitan Flood Control District
Kansas Department of Health & Environment
Kentucky Department of Environmental Protection
Ohio River Valley Sanitation Commission
Urban Drainage & Flood Control District, CO

CORPORATE

ADS Environmental Services
Alan Plummer & Associates
Alden Research Laboratory Inc.
Alpine Technology Inc.
Aqua-Aerobic Systems Inc.
Aquateam–Norwegian Water Technology Centre A/S
ARCADIS
Associated Engineering
Black & Veatch
Boyle Engineering Corporation
Brown & Caldwell
Burgess & Niple, Ltd.
Burns & McDonnell
CABE Associates Inc.
The Cadmus Group
Camp Dresser & McKee Inc.
Carollo Engineers Inc.
Carpenter Environmental Associates Inc.
CDS Technologies Inc.
CET Engineering Services
Chemtrac Systems Inc.
CH2M HILL
Conestoga-Rovers & Associates
CONTECH Stormwater Solutions
D&B/Guarino Engineers, LLC
Damon S. Williams Associates, LLC
Earth Tech Inc.
Ecovation
EMA Inc.
Environ/The ADVENT Group, Inc.
Fay, Spofford, & Thorndike Inc.
Freese & Nichols Inc.
ftn Associates Inc.
Fuss & O'Neill Inc.
Gannett Fleming Inc.
Geosyntec Consultants
GHD
Golder Associates Ltd.
Greeley and Hansen LLC
Hazen & Sawyer, P.C.
HDR Engineering Inc.
HNTB Corporation
Hydromantis Inc.
HydroQual Inc.
Infilco Degremont Inc.
Jacobson Helgoth Consultants Inc.
Jason Consultants LLC Inc.
Jordan, Jones, & Goulding Inc.
KCI Technologies Inc.
Kelly & Weaver, P.C.
Kennedy/Jenks Consultants
KMK Consultants
Komline Sanderson Engineering Corporation
Limno-Tech Inc.
Lombardo Associates Inc.
The Low Impact Development Center Inc.
Malcolm Pirnie Inc.
Material Matters
McKim & Creed
MEC Analytical Systems, Inc.
Metcalf & Eddy Inc.
Monteco Corporation
MPR Engineering Corporation, Inc.
MWH
O'Brien & Gere Engineers Inc.
Odor & Corrosion Technology Consultants Inc.
Oscar Larson & Associates
PA Government Services Inc.
Parametrix Inc.
Parsons
Post, Buckley, Schuh & Jernigan
RMC Water & Environment
R.M. Towill Corporation
Ross & Associates Ltd.
Rothberg, Tamburini & Windsor, Inc.
SAIC
Siemens Water Technologies
Stantec Consulting Inc.
Stearns & Wheler, LLC
Stone Environmental Inc.
Stratus Consulting Inc.
Synagro Technologies Inc.
Tetra Tech Inc.
Trojan Technologies Inc.
Trussell Technologies, Inc.
URS Corporation
Wade-Trim Inc.
Westin Engineering Inc.
Weston Solutions Inc.
Woodard & Curran
Zenon Environmental Inc.
Zoeller Pump Company

INDUSTRY

American Electric Power
American Water
ChevronTexaco Energy Research & Technology Company
The Coca-Cola Company
Dow Chemical Company
DuPont Company
Eastman Chemical Company
Eastman Kodak Company
Eli Lilly & Company
Merck & Company Inc.
Premier Chemicals LLC
Procter & Gamble Company
RWE Thames Water Plc
Severn Trent Services Inc.
Suez Environnment
United Water Services LLC

Note: List as of 9/1/06

IWA Publishing's authorised EU representative for General Product Safety Regulations is Diane D'Arras, 15 rue Duret, 75116 Paris, France, e-mail: safety@iwap.co.uk.

Printed and bound by CPI Group (UK) Ltd, Croydon, CR0 4YY
12/05/2026
02108893-0005